Springer Textbooks in Earth Sciences, Geography and Environment

The Springer Textbooks series publishes a broad portfolio of textbooks on Earth Sciences, Geography and Environmental Science. Springer textbooks provide comprehensive introductions as well as in-depth knowledge for advanced studies. A clear, reader-friendly layout and features such as end-of-chapter summaries, work examples, exercises, and glossaries help the reader to access the subject. Springer textbooks are essential for students, researchers and applied scientists.

Branislava Lalic · Josef Eitzinger · Thomas Foken ·
Tamás Weidinger
Editors

Micrometeorological Measurements

An Introduction for Beginners

COST
EUROPEAN COOPERATION
IN SCIENCE & TECHNOLOGY

Funded by
the European Union

Springer

Editors
Branislava Lalic
University of Novi Sad
Novi Sad, Serbia

Josef Eitzinger
BOKU University
Vienna, Austria

Thomas Foken
University of Bayreuth
Bayreuth, Germany

Tamás Weidinger
Eötvös Loránd University
Budapest, Hungary

ISSN 2510-1307 ISSN 2510-1315 (electronic)
Springer Textbooks in Earth Sciences, Geography and Environment
ISBN 978-3-032-03883-8 ISBN 978-3-032-03884-5 (eBook)
https://doi.org/10.1007/978-3-032-03884-5
Earth Science Institute of Slovak Academy of Science COST CA20108

This Springer imprint is published by the registered company Springer Nature Switzerland AG
The registered company address is: Gewerbestrasse 11, 6330 Cham, Switzerland

Preface

Micrometeorological Measurements—An Introduction for Beginners is designed and written as a guideline for individuals who need to organize and conduct micrometeorological measurements but lack the necessary knowledge and skills, and don't have the time to acquire them through regular training. It caters to beginners who have a basic understanding of physical processes and are at the start of their professional careers, as well as professionals who are familiar with meteorological elements but not deeply versed in physical processes and meteorological instrumentation.

Before you start reading, take a moment to reconsider the goal of your experiment and the role of micrometeorological measurements in achieving that goal. Decide whether you want to measure the meteorological conditions of a specific location regardless of environmental changes, focus on a specific environment, or register the effects of processes expected at a specific location. Once you are clear on your goal and the means to achieve it, this book will guide you—from sensor and location selection, through measurement and data collection, to quality control and filling of potential data gaps.

▶ Chapter 1 provides a brief physical introduction to governing processes and their effects. Understanding the causes and consequences of atmosphere-biosphere interaction is essential for designing reliable experiments. ▶ Chapter 2 helps improve the representativeness of your measurements and guides you in sensor selection (type, inertia, principle of work), data collection, and transfer methodology. ▶ Chapter 3 describes the measuring devices used for each meteorological element discussed, with selection tailored to the application objective. ▶ Chapter 4 introduces best practices and methods for quality assurance and control of measured micrometeorological data, vital to correctly identifying, addressing, and correcting errors that may occur during data collection, processing, and analysis.

Filled with first-hand experience in micrometeorological measurements, this book offers unique value. By the end, you will be equipped to make informed decisions about the design of micrometeorological experiments, instrumentation, and data quality assurance and control—while addressing common measurement challenges.

Branislava Lalic
Novi Sad, Serbia

Josef Eitzinger
Vienna, Austria

Thomas Foken
Bayreuth, Germany

Tamás Weidinger
Budapest, Hungary

Acknowledgements

This publication is based on work carried out in the framework of the COST Action FAIRNESS (FAIR NEtwork of micrometeorological measurements), CA20108, supported by COST (European Cooperation in Science and Technology).

COST (European Cooperation in Science and Technology) is a funding agency for research and innovation networks. Our Actions help connect research initiatives across Europe and enable scientists to grow their ideas by sharing them with their peers. This boosts their research, career and innovation.

The COST Association, an International non-profit Association, funds and coordinates scientific and technological research across Europe and internationally. The COST Association has currently 41 members and 1 cooperating member (► www.cost.eu)

EUROPEAN COOPERATION IN SCIENCE & TECHNOLOGY

Funded by the European Union

FAIRNESS

Contents

About the Editors

Branislava Lalic is a professor of Meteorology and Biophysics at Faculty of Agriculture, UNS, Novi Sad (Serbia). Decades of experience in teaching micrometeorological measurements, instrumentation and data management, particularly to agronomy students and early-career researchers inspired this beginner-oriented guide.

Josef Eitzinger is professor for Agrometeorology at the University of Natural Resources and Life Sciences, BOKU, Vienna (Austria). His long-time expertise and research focus includes the processes of soil-crop-atmosphere interactions, crop modelling, crop-water balance, drought and cropping risks monitoring, agrometeorological measurements and climate change impact assessments on crop production.

Thomas Foken is a retired professor of micrometeorology at the University of Bayreuth (Germany) with many years of experience in research and teaching on the interaction between the Earth's surface and the atmosphere. He previously headed departments at the Meteorological Observatories in Potsdam and Lindenberg. His work on experimental meteorology has been honoured internationally.

Tamás Weidinger is a retired Associate Professor at Eötvös Loránd University (ELTE, Budapest, Hungary), with a Ph.D. in meteorology. His main research topic is surface-biosphere-atmosphere exchange processes including both measurements and modelling. He is an editor of Theoretical and Applied Climatology and Editorial Board member of Időjárás (Quarterly Journal of the HungaroMet).

Contributors

Peter Domonkos is a Hungarian climatologist living in Spain. Ph.D. (Kossuth Lajos University, Debrecen, Hungary, 1999), expert on statistical climatology, analysis of extreme climatic events, data quality control and time series homogenization. Software developer and author of a large number of scientific papers.

Zoltán Istenes is an Associate Professor at Eötvös Loránd University (ELTE, Budapest, Hungary), specialized in robotics, intelligent systems, and sensor technologies. His research includes the development of advanced meteorological data acquisition platforms, ranging from ground-based weather stations to unmanned aerial vehicles (UAVs) equipped with environmental and atmospheric sensors for real-time, in-situ observations.

Britta Jänicke is professor of Environmental Meteorology at the University of Kassel. Her research focuses on urban climate and human biometeorology, with expertise in multi-scale observations and simulations. With a background in Urban Ecology and Landscape Planning, she has experience linking urban climatology and planning to support climate-resilient cities.

Ivan Koči is MS Certified professional, responsible for PIS agrometeorological network (Serbia) and its data assimilation and analysis.

Erich Mursch-Radlgruber is retired professor for boundary layer meteorology at the University of Natural Resources and Life Sciences, BOKU, Vienna, Austria. His expertise and research focus includes micro- and mesoscale meteorological analysis, including related measurement and modelling techniques.

Mina Petrić is a researcher at Avia-GIS, Belgium, since 2016. She is a physicist specialised in meteorology with a focus on mathematical vector population dynamics modelling and how population seasonality and abundance is affected by weather patterns, changing climatic conditions and extreme events.

Thomas Vergauwen is specialized in high-resolution numerical weather forecasting and non-traditional observations. His primary focus is on verification techniques using crowdsourced observations and maintaining the MetObs-toolkit. With a background in physics and meteorology, he is affiliated with the Royal Meteorological Institute of Belgium (RMI) and Ghent University.

Introduction to Micrometeorological Measurements

Thomas Foken, Tamás Weidinger, Branislava Lalic and Josef Eitzinger

B. Lalic et al. (eds.), *Micrometeorological Measurements*,
Springer Textbooks in Earth Sciences, Geography and Environment,
https://doi.org/10.1007/978-3-032-03884-5_1

The introductory chapter describes the necessary basic knowledge for (micro)meteorological measurements in the context of different applications and for the understanding of the atmospheric boundary layer and meteorological scales in which frame measurements are typically carried out. Only SI units (SI 2019) are used throughout the book (see Appendix 1). The physical quantities used are related to the current temperature scale (ITS-90) according to Foken et al. (2021a). Within these Guidelines, we use the term **variable** for quantities that change over time and/or space and describe the system state and the term **parameter** for constant or well-defined characteristics of a particular system and, if not constant, their behavior under changing conditions is known.

1.1 The Basic Scientific Background to Be Known

The interaction of the atmosphere with the other geospheres is of decisive importance for the Earth's climate and its change. In respect of time scales, climate is defined as the appearance of the—in short-time changing—atmospheric conditions (expressed as "weather") over longer periods of time (e.g., a minimum of a 30-year period defined by WMO) in statistical or descriptive terms. Directly at the Earth's surface, the crucial exchange of momentum (pushing force on/of air mass movement), energy (such as warming/cooling of air mass at the surfaces and radiation absorption/emission), and matter (such as water and various gases and aerosol particles) with the atmosphere takes place. These processes are driving the atmospheric circulation systems covering the entire Earth.

- **Weather** represents short-term and small-scale changes in the atmosphere.
- **Climate** is the synthesis of weather over a period long enough to establish its statistical characteristics (mean values, variances, probabilities of extreme events, etc.) (WMO 1979).

The climate system can only be described correctly in its complexity if the necessary process understanding for these exchanges between Earth surface and atmosphere is available, which proves to be much more complex than what simple graphics in climate textbooks show.

The basic prerequisite and driving force for the formation of the climate and all weather processes is the amount of energy which the Earth receives from the sun through shortwave radiation. Mainly at the Earth's surface (beside some atmospheric contribution) the absorbed part of shortwave radiation is converted either in longwave (thermal) terrestrial radiation or in turbulent energy fluxes (sensible heat flux and latent heat flux), ground heat flux and any heat storage. The partitioning and balance in these different energy fluxes is highly variable in time and space, depending on atmospheric, weather, and surface conditions.

The exchange of momentum, energy, and matter between the Earth's surface and the atmosphere also has very practical consequences. Modifications of surface properties, such as natural vegetation and soil, or man-made ones, such as soil sealing, can create climate types or characteristics even on a small (micro, local) scale. In addition to the increase of greenhouse gases in the atmosphere by fossil energy burning and other human activities, land-use changes are therefore an additional component of global and especially local climate change through various effects on surface exchange processes. On a larger scale compared to natural conditions in a region, it can, for example, contribute to increased desertification or changes in the water balance. Man-made land-use changes, however, can have deliberate (intended, optimizing) impacts on microclimate as well as unintended ones (such as overheating of cities).

1.1.1 Short- and Longwave Radiation Characteristics, Fluxes, and Balance

A large part of the sun's radiation energy is emitted in visible light due to its high surface temperature of about 6000 K. In addition, there are components in ultraviolet and infrared light, so that the wavelength range reaching the Earth's surface is 0.29–3 μm and is referred to shortwave or solar (extraterrestrial) radiation. High-energy radiation with very short wavelengths, such as ultraviolet radiation, is mostly absorbed by the ozone layer of the stratosphere. Longwave radiation which is emitted only from the Earth's surfaces and atmosphere components is also affected by atmospheric absorption. Shortwave radiation is absorbed only in some wavelength ranges. An overview of the absorption ranges by various gases within the atmosphere in the short- and longwave radiation range is shown in ◘ Fig. 1.1.

The sun emits an energy of 4×10^{20} MW. At a mean distance between the Earth and the sun of 149.6×10^6 km, the Earth then receives about 1361 W m^{-2} at the upper limit of the atmosphere (referred to as the solar constant at perpendicular incidence, Kopp and Lean 2011) with a standard uncertainty of 0.5 W m^{-2}. Since the sun does not shine all day long and the angle of incidence varies, the global average irradiance at the upper atmosphere is 340 W m^{-2}. In the mid-latitudes, almost 500 W m^{-2} of shortwave radiation reach us as in the daily average during summer in the outer atmosphere and in winter only 100 W m^{-2}. These are exactly energy flux densities, i.e., amounts of energy in joules per unit area and time, which only becomes clear by converting $1\ \mathrm{W\ m^{-2}} = 1\ \mathrm{J\ m^{-2}\ s^{-1}}$.

> **Note 1.1** There are different values of the solar constant reported in the scientific literature, ranging from 1361 to 1367 W m^{-2}. The reason is the differences in the applied sensors and the measurement techniques and not a change in solar radiation. This book uses the value determined using state-of-the-art technology according to Kopp and Lean (2011) and the radiation distribution in the Earth's atmosphere derived from this.

During transmission of the shortwave solar radiation through the atmosphere, various atmospheric gases absorb specific wavelengths of the solar spectrum, changing the spectral wavelength distribution of the sunlight (◘ Fig. 1.2) as well as reducing the amount

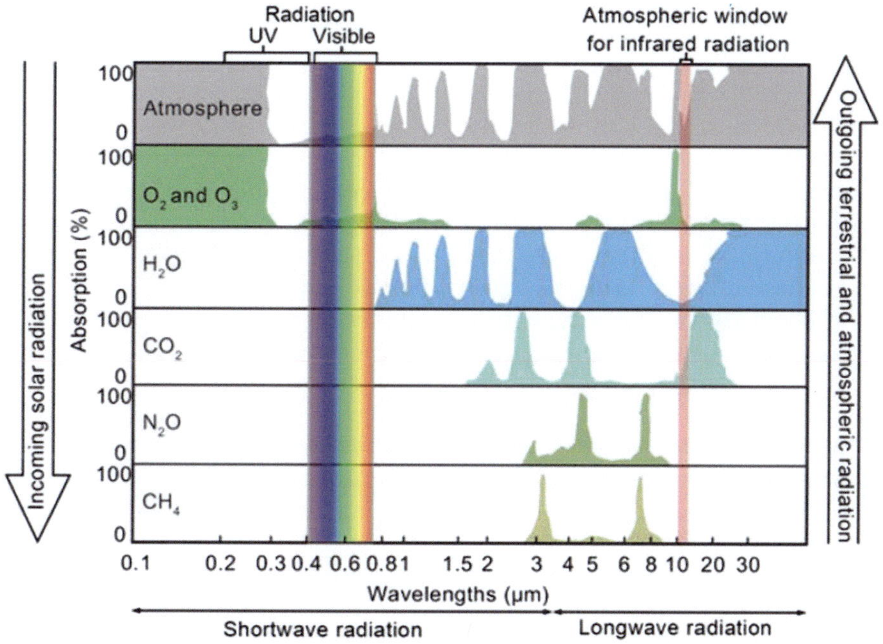

◘ **Fig. 1.1** Major absorption bands of the atmosphere in the short- and longwave radiation and the contribution of various gases to it, after Lalic et al (2018), © Authors

1

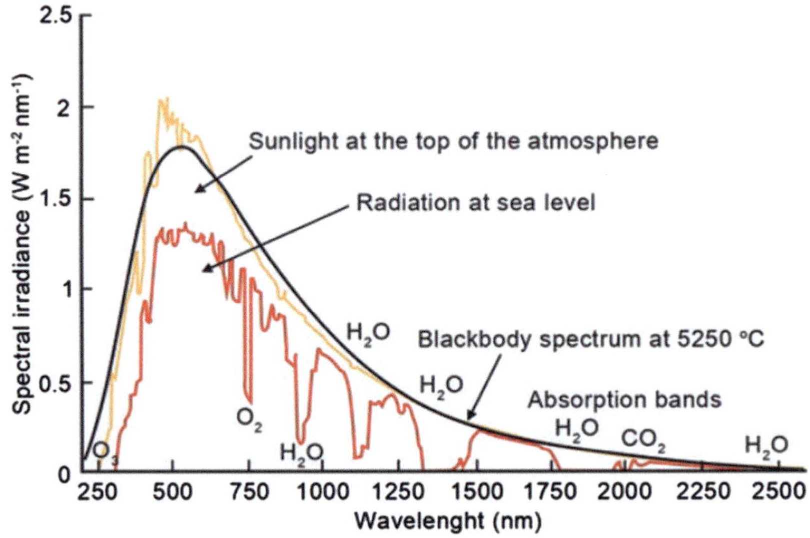

Fig. 1.2 Shortwave solar (or global) radiation spectra at the top and bottom of the atmosphere and related absorption ranges caused by atmospheric gases, after Lalic et al. (2018). © Authors

of solar energy reaching the Earth's surface. For example, the daily maximum global radiation at the ground in the mid-latitudes of the Northern Hemisphere reaches about 1000 W m^{-2} on a clear summer day in June/July. Similarly, the sunlight energy and its spectrum are further significantly reduced or modified when transmitted through other materials such as (green) canopies, plant leaves, and water bodies.

The major part of the energy reaches the Earth's surface through direct shortwave solar radiation (the solar beam). Diffuse shortwave radiation (scattered in the atmosphere) reaches the Earth's surface under cloud cover or in a shadow during sunshine (i.e., the radiation portion without direct sunlight), which contains only about 20% of the energy of the total shortwave solar radiation (with modified spectral distribution).

The fraction of the shortwave radiation absorbed by atmospheric gases (especially ozone and water vapor) contribute beside absorbing aerosols to a warming effect at different atmospheric layers. Stratospheric ozone causes the warming of the stratosphere by the absorbed UV radiation. Clouds significantly modify the shortwave radiation fluxes in the atmosphere (by reflection) and reduce absorption and related warming effects at the earth surface and the lower atmosphere during daytime (consider the differences to the co-existing thermal radiation balance effects further below). However, the part of the shortwave radiation reaching the Earth's surface is partly absorbed and partly reflected by all existing surfaces depending on their mostly color-dependent albedo and heats them up by the absorbed part. Water and moist surfaces can transmit and thus store a particularly large amount of energy due to their relatively high absorption rates as well as their high heat capacity and heat conduction (see ▶ Sect. 2.2). Other surfaces, such as sand, become significantly hotter at the surface, also in spite of high albedo (reflection), because heat capacity and heat conduction are low, so that even a small amount of radiation is sufficient to heat up a thin surface layer vigorously.

However, at the Earth's surface the actual sun beam angle is the main parameter defining how much solar energy can be absorbed by a unit of surface area. The inclination of the Earth's axis by about 23.5° in respect of the orbital plane of the Earth around the sun (ecliptic) leads to high sun elevations and stronger solar radiation in summer and low sun elevations in winter. This is the main cause of the typical seasonal climate of mid-to-high latitudes (Fig. 1.3a). The exposure of a surface to the sun is modified addition-

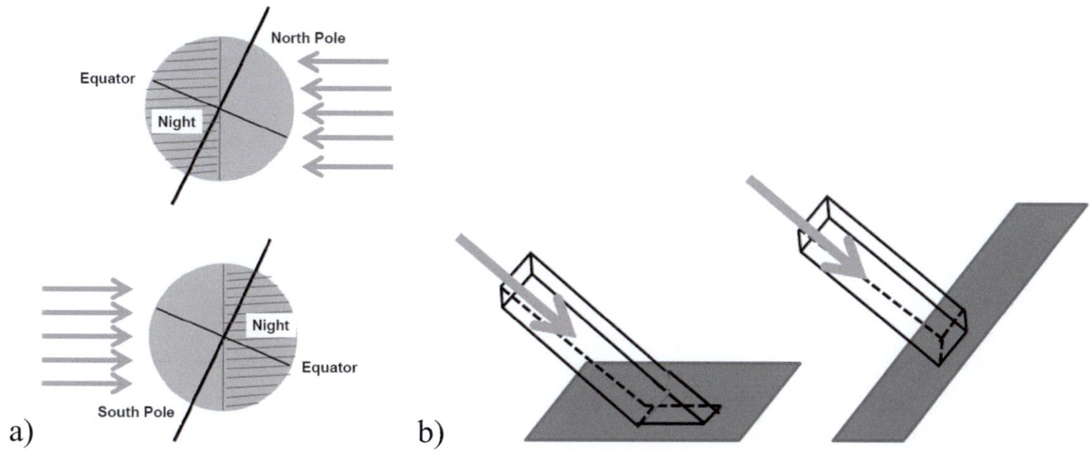

Fig. 1.3 **a** Position of the Earth to the sun at the (northern) beginning of summer (top) and at the (northern) beginning of winter (bottom); **b** schematic representation of perpendicular (right) and oblique (left) incidence of the sun's rays on a surface, after Foken (2013) © Author, CC BY 4.0

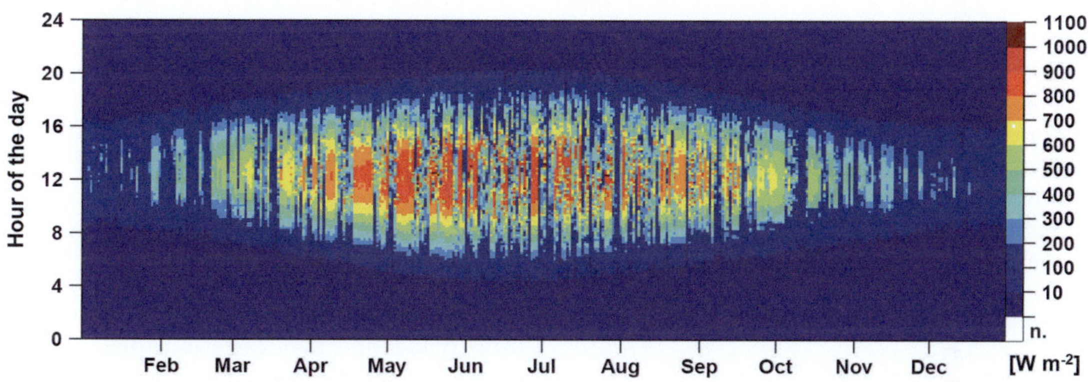

Fig. 1.4 Hovmøller plot of global radiation in W m^{-2} for the year 2011 in the Ecological-Botanical Garden of the University of Bayreuth, Germany, in UTC. The picture for the reflected radiation looks similar; the values are only smaller according to the albedo (Table 1.1), after Foken (2013) © Author, CC BY 4.0. The Hovmøller plot is a particularly descriptive representation of three parameters (Persson 2017), usually with the months on the x-axis and the time of day on the y-axis (latitude is also common). The third parameter is shown in color

ally by inclined surfaces (slope and slope orientation), which plays a special role in the local climate (Fig. 1.3b).

Keeping in mind that the sun beam angle is changing in the daily as well as seasonal course depending on the latitude, Fig. 1.4 shows an example of measured global radiation for a specific year from a Central European site. From that example, it can be seen that the annual course and variation of the global radiation are affected besides day length and sun beam angle additionally by the specific weather conditions and changing cloudiness.

Moreover, in the mid-latitude spring, particularly high radiation values are often recorded on cloudless days (see April–May in Fig. 1.4), since the air masses are usually still cool and have a high permeability for shortwave radiation due to the low water vapor and aerosol content (due to low evaporation rates and convection in the winter half-year). This phenomenon is not pronounced in autumn despite Indian summer. In summer, due to many convective clouds or unstable weather conditions, a rather mixed radiation climate is encountered.

1

> **Note 1.2** Because of the importance of global radiation for all energy conversions at the Earth's surface, it should be measured in all micrometeorological studies or measured data used from a nearby standardized weather station.

The ratio of reflected and incident shortwave radiation is called albedo and is usually expressed as a percentage (◘ Table 1.1). The higher albedo so brighter is the surface, so that almost 100% is achieved with freshly fallen snow. Sand surfaces are at 30–40% and water is lowest at well under 10% with decreasing sun angle. Most photosynthetic active (green) plant surfaces or canopies are around 15–25%.

All surfaces of any material emit thermal radiation in a certain wavelength range depending on its surface temperature. This is also true for the Earth. At typical temperatures at the Earth's surface, terrestrial longwave radiation is emitted in a wavelength range from about 3–100 µm with a maximum at about 10–12 µm. The emitted energy is determined according to the Stefan-Boltzmann law, whereby a surface-dependent emission coefficient (◘ Table 1.2) must be used.

◘ **Table 1.1** Albedo of different surfaces (Foken and Mauder 2024)

Surface	Albedo
Clean snow	75–98
Grey soil, dry	25–30
Grey soil, wet	10–12
White sand	34–40
Wheat	10–25
Oaks	18–20
Pine	18
Water, rough, solar angle 90°	14
Water, rough, solar angle 30°	13

◘ **Table 1.2** Infrared emissivity of different main surface types (Foken and Mauder 2024)

Surface	Emissivity
Water	0.960
Fresh snow	0.986
Coniferous needles	0.971
Dry fine sand	0.949
Wet fine sand	0.962
Thick green grass	0.986

> **Note 1.3** Stefan-Boltzmann law
>
> $$I = \varepsilon_{IR} \sigma_{SB} T^4 \quad \text{(N1.1)}$$
>
> with the longwave radiation I, the Stefan-Boltzmann constant $\sigma_{SB} = 5.67 \times 10^{-8}$ W m^{-2} K^{-4}, the emission coefficient ε_{IR}, and absolute temperature T in kelvin.

Unlike shortwave radiation, longwave radiation can contribute to the heating of any solid material surfaces as well as the air. Its wavelengths can be absorbed—besides solid aerosol particles such as dust in the air—by non-symmetric molecules such as those of water vapor, carbon dioxide, methane, nitrous oxide, and tropospheric ozone, the so-called greenhouse gases. These gases heat up slightly and re-radiate the absorbed energy in all directions based on their temperature. This process is known as re-emission of longwave radiation.

In this way, the atmosphere receives back a large part of the emitted longwave radiation and the longwave radiation flux upwards is reduced, so that the stratosphere cools down, while the counter-radiation directed downwards allows additional heating in the troposphere. Since the two spheres are poorly coupled, little compensation occurs. While the longwave radiation coming from the Earth's surface is called upwelling longwave radiation, the opposite is called down-

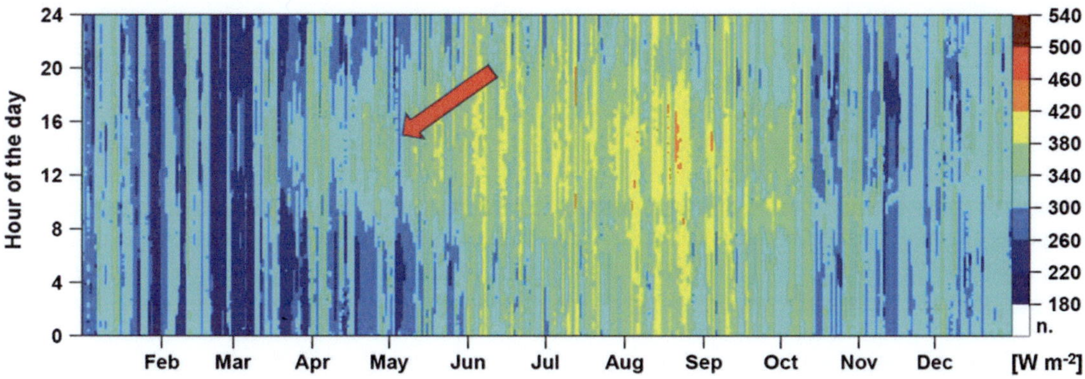

Fig. 1.5 Hovmøller plot of the downwelling longwave radiation in W m^{-2} for the year 2011 in the Ecological-Botanical Garden of the University of Bayreuth (Germany), in UTC, after Foken (2013) © Author, CC BY 4.0

welling longwave radiation (source is the atmosphere).

An annual variation of the downwelling longwave radiation for a Central European location is shown in Fig. 1.5, where a late frost event for that year in May 2011 is also visible (red arrow). Especially in winter and early spring, often such periods with particularly low downwelling longwave radiation can occur due to low aerosol concentrations in the local cold air masses and in combination with cloudless conditions and low air humidity in the atmosphere. This is mostly caused by synoptic reasons (cold air mass inflow) outweighing the seasonal effects of increasing day length in spring leading to higher energy input from the sun on the Earth's surface. Such low downwelling longwave radiation increases the negative thermal surface radiation balance during night significantly and allows the surface to cool down particularly strongly with a corresponding danger of frost.

A stronger diurnal cycle is noticeable during the warmer summer period. During the day, the atmosphere warms up more significantly, because more aerosols and water vapor are introduced in the atmosphere due to convective processes.

The annual cycle of the upwelling longwave radiation is shown in Fig. 1.6. It is very similar to the global radiation, but the surface temperatures show muted dynamics due to the ground heat flux dynamics, al-

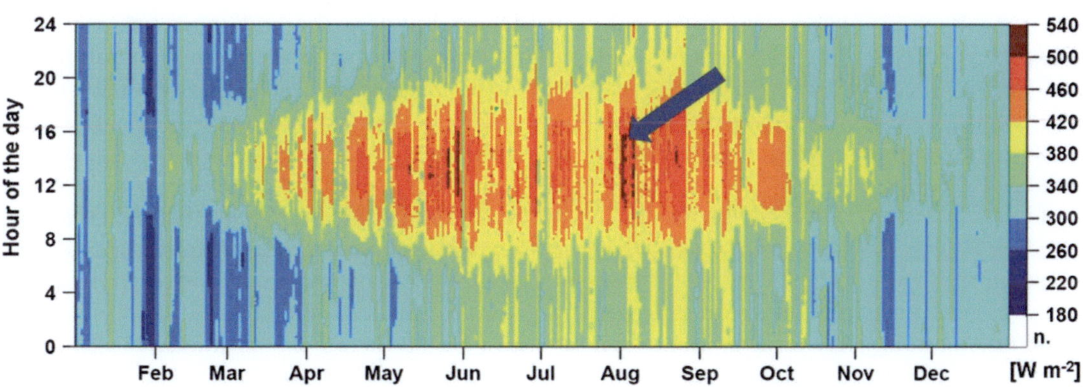

Fig. 1.6 Hovmøller plot of the upwelling longwave radiation in W m^{-2} for the year 2011 in the Ecological-Botanical Garden of the University of Bayreuth, Germany, in UTC, after Foken (2013) © Author, CC BY 4.0

bedo effects, and latent heat exchange, depending on the type of surface condition. For example, at the beginning of August a strong drought leads obviously to less evaporative surface cooling (as well as potentially to lower soil heat conductivity) causing heating up of soil surface expressed as higher upwelling thermal radiation (blue arrow).

The sum (balance) of the incident shortwave global radiation, which consists as mentioned of the diffuse and the direct solar radiation, and the reflected shortwave radiation from a surface is called **shortwave radiation balance**. Together with the **longwave radiation balance** (the longwave terrestrial radiation emitted from the surface and the amount absorbed by the surface from the atmosphere), we receive the **total radiation balance or net radiation** (see Note 1.4). The reference surface (e.g., soil surface or a plant canopy) of the shortwave and longwave radiation balance is also called the **energy exchange surface** (see also Energy Balance in ▸ Sect. 1.1.4).

Note 1.4 Radiation balance equation:

$$Q_s^* = K\uparrow + K\downarrow + I\uparrow + I\downarrow \qquad \text{(N1.2)}$$

with the radiation balance Q_s^*, the downwelling shortwave global radiation $K\downarrow$, the reflected shortwave radiation $K\uparrow$, the upwelling longwave terrestrial radiation emitted from the surface $I\uparrow$, and the downwelling longwave radiation from the atmosphere $I\downarrow$. The radiation balance is variable over time (especially depending on weather conditions, seasonal or day and night cycle, changing surface conditions) and can be measured or expressed in different time units or referred to as different spatial units.

Radiation fluxes as well as the turbulent fluxes—as discussed below—are expressed in micrometeorology as energy amounts. On average, the Earth's surface receives more radiation energy than it emits; however, the direction of fluxes can change over time (e.g., due to seasonal and day-night cycle). For common understanding of the direction of energy fluxes from and to a surface (see Note 1.5 for further information), following sign convention applies in the following text:

Energy fluxes are set to positive values if they dissipate energy from the energy exchange surface (i.e., the Earth's surface) into the atmosphere or into the ground (or mass), and they are set as negative values and vice versa. A corresponding diurnal cycle is shown as an example for a cloudless day in ◘ Fig. 1.7.

The Earth's surface receives 103 W m^{-2} as the sum of all radiative fluxes (= radiation balance or net radiation) on global average, but it can vary largely in respect of location, daytime, season, or weather conditions.

The radiation balance (Note 1.4) has to be specified for a certain space and time unit and explains the amount of energy which is available for further energy exchange processes (called also "**available energy**"; see Note 1.6) which happen at the same time at the same place at the relevant surface. These energy exchange processes at any surface are described by **the energy balance equation** (Note 1.7) which explains for which processes and to which parts the available energy from the radiation balance is actually used.

In other words, to balance the energy balance terms (see Note 1.5 and ▸ Sect. 1.1.4), the energy exchange surface releases energy through turbulent fluxes (heat and water) into the atmosphere (which is a mass transport of gases) and into the ground (which is represented as a heat storage in the ground). Depending on the radiation balance, also the energy balance shows high variability in time and space and all energy balance terms can be positive or negative to certain levels. For example, the ground heat flux in the seasonal European climates is overall positive in spring (in phases where the soil is warming up) and becomes overall negative in fall/winter (in phases when soil is cooling down). Due to this seasonal climatic driven cycle, it can be neglected in the annual energy balance. Similar behavior of ground heat flux is observable in the diurnal (day and night) cycle, but much stronger influenced by short-term changing weather conditions, which

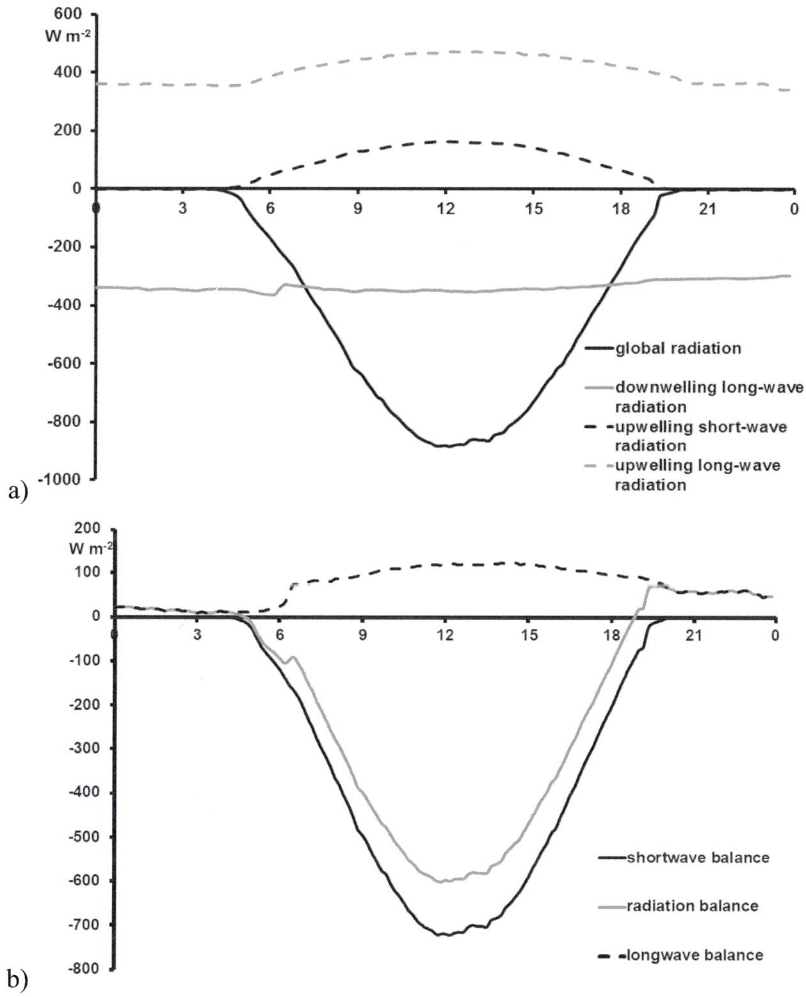

Fig. 1.7 **a** Diurnal variation of the components of the radiation balance and **b** the short- and longwave radiation balance on 05.24.2012 in the Ecological-Botanical Garden of the University of Bayreuth (in CET, sign convention; see Note 1.5), after Foken (2013) © Author, CC BY 4.0

does often not allow a neglect for daily-based analysis.

Note 1.5 Sign conventions for radiation and energy fluxes:
The choice of the sign for energy and radiation fluxes is independent of the actual direction of the energy flux. It depends on how one writes the energy balance equation or how one defines the gradient for energy fluxes. Two definitions are common if one writes all terms of the energy balance equation as a sum on one side of the equation: In general meteorology, the global radiation is assumed to be positive, and it follows that during the day the sensible and latent heat flux and the ground heat flux are negative. In micrometeorology (as in this book), because of its origin in agricultural meteorology, evaporation (latent heat flux) is assumed to be positive during the day, so the same is true for sensible heat flux and ground heat flux, but then global radiation as the energy input driving these fluxes has to be negative to balance the energy budget.

1

Note 1.6 Available energy:
For many micrometeorological measurements, it is important to know the available energy for the energy balance processes due to the radiation balance. For this purpose, net radiometers that directly measure the radiation balance can be used. However, net radiometers that measure all four radiation components separately are more accurate (see ▶ Sect. 3.5).

1.1.2 Turbulent Fluxes

A special property of atmospheric flow is atmospheric turbulence. Here, individual air parcels (much larger than molecules: turbulence elements, turbulence vortices) move irregularly and randomly around a mean state. These turbulence elements differ with respect to their density (temperature- and humidity-dependent) and their motion.

By structuring the atmosphere into turbulence elements, the **transport of momentum** (product of mass and velocity, a measure of friction), heat, and gases (with another word: properties) becomes much more efficient than in molecular conduction processes. ◘ Figure 1.8a shows a conceptual picture for heat and water vapour flux. Warm air normally rises because of lower density and cold turbulence elements descend. If, due to friction on the surface, this is further supported by corresponding vertical velocities, we no longer speak of thermal turbulence, but of thermal-dynamic turbulence. Thus, it is also possible that, to a certain extent, warm turbulence elements with a negative vertical velocity can move toward the ground and cold ones with a positive vertical velocity into higher air layers. The exchange of heat is called **sensible heat flux**.

The sensible heat flux is responsible for the heating of the air after sunrise, while the nocturnal cooling is caused by negative longwave radiation balance, resulting in a negative sensible heat flux. The sign convention is analogous to the convention for radiation fluxes (see Note 1.5). The surface heated by the shortwave radiation transfers its heat in the lowest millimeter first from molecule to molecule (molecular conduction), but already immediately above that to turbulence elements, the smallest of which have a size of centimeters. Smaller turbulence elements unite to larger ones, but in larger one's smaller ones are still present. These can move relatively quickly away from the surface and make the turbulent exchange much more effective than the molecular conduction directly above the surface. At 2 m height above ground, several decimeters are already the typical size of these eddies and with increasing height they become larger and larger. The sensible heat flux is responsible for the heating of the air up to several 100 m height in the daytime.

The situation is similar with evaporation, i.e., the flux of water vapor or latent heat flux (◘ Fig. 1.8b). The exchange of moisture (note: moist air is lighter (has less specific weight)) than dry air at the same temperature, since water vapor has a lower molecular weight of 18 g mol^{-1} compared to dry air of 29 g mol^{-1}) and of other gaseous substances takes place in the same way.

Here, too, water vapor is transported by molecular diffusion above the evaporating surface (e.g., soil, water or leaves) in the first millimeter. Above this, turbulence elements are formed with more or less content of water vapor, i.e., they are lighter or heavier than their surroundings. As a first approximation in the surface layer, the water vapor is transported by the given eddy in the same way as the sensible heat. The water vapor flux is a passive characteristic of turbulence in this sense. That is, water vapor "just suffers" the turbulent transport, but does not shape its development.

Again, the transport becomes very effective together with the vertical momentum component of the wind. Evaporation is also called **latent heat flux**, because a lot of energy ("latent energy") is needed for the transition of water into water vapor, which is only released when the water vapor condenses in the cloud. Water vapor thus exerts a latent energy transport. Evaporation of vegetated surfaces is often referred to as **evapotranspiration(evaporation + transpiration)**. **Transpira-**

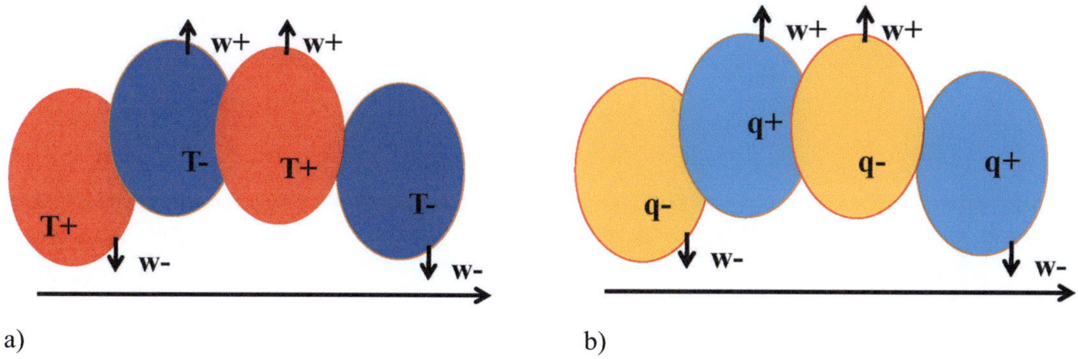

Fig. 1.8 **a** Schematic representation of warm (T+) and cold (T−) turbulence elements. Through the connection with an upward directed vertical wind component (w+) or a downwardly directed vertical wind component (w−), the sensible heat flux is generated in the turbulent wind field, **b** analogous conditions for humid (q+) and dry (q−) air for water vapor transport (evaporation, latent heat flux), after Foken (2013)

tion is the release of water from plant leaves through their stomata and **evaporation** is the release of water from all wet surfaces except the release through plant stomata path. For example, after rainfall, water on leaves evaporate first (called interception).

1.1.3 Ground Heat Flux

As a part of the energy balance terms (Note 1.7), the ground heat flux quantifies how much energy in transported into or out of a solid material (such as the soil surface in natural environment) in a unit of time and space. As already shown in ▶ Sect. 1.1.1, the energy exchange surface, commonly the Earth's surface (soil, plants, urban areas), heats up during the day due to the incoming shortwave radiation and cools down during the night due to negative thermal (longwave) radiation balance. Usually, only a small part of this energy is conducted into the soil or any other surface mass, where it is temporarily stored.

Shading of the soil or ground (e.g., by a plant canopy) strongly reduces soil heat flux. Further, the relatively low efficiency of the ground heat flux is related to the largely only molecular heat conduction in the soil, expressed by the molecular heat conduction coefficient, which depends on the soil particle composition and water/air content.

For meteorological or climatological large-scale applications (such as weather forecast or climate models), the small-scale heterogeneity of the soil and surface characteristics determining ground heat flux can hardly be taken in account. In the daily cycle, however, only the uppermost soil layers are warmed up (Fig. 1.9) and the summertime seasonal warming usually extends less than about 20 m into the soil. The annual cycle of the ground temperatures over one year is shown in Fig. 1.10. From this, a reversal of the gradient of the ground temperature with the depth and thus the direction of the ground heat flux can be determined, for example, at about the beginning of April and the end of September in the Central European climate. From Fig. 1.9 it can be seen that the ground heat flux near a soil surface (not shaded), even on a radiative, hot day in summer during the midday hours, is relatively small compared to the other energy fluxes, at about 50–100 W m^{-2}. This amount of energy predominantly heats the top 10–20 cm soil layer. How much energy is available for heating the soil and for other energy fluxes depends on the radiation balance at the soil surface. How much energy the soil can store or transfer to deeper layers depends on the heat capacity and thermal conductivity of the soil components and the dynamically changing soil water content.

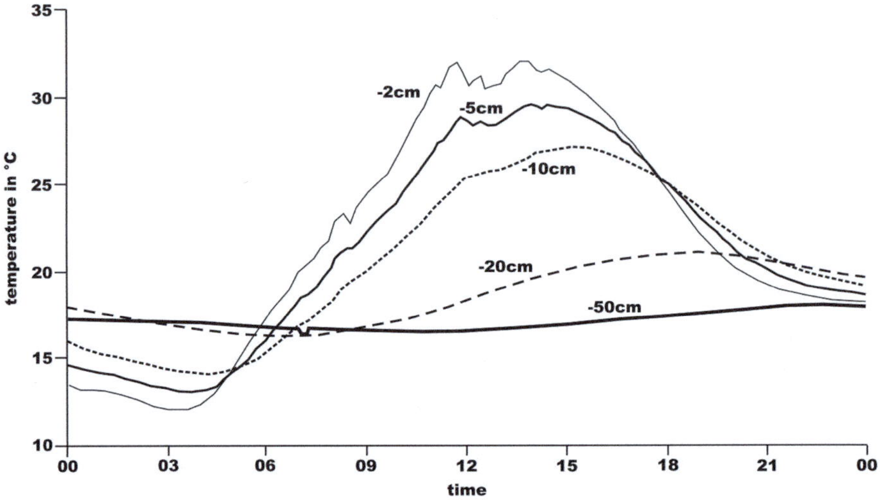

Fig. 1.9 Diurnal variation of soil temperatures at different depths on 06.05.1998, measured by the University of Bayreuth during the experiment LITFASS-98 (fallow) near the Boundary Layer Measurement Field of the Meteorological Observatory Lindenberg, German Meteorological Service (in UTC, from 12 to 14 o'clock passage of high cloud cover) according to Foken and Mauder (2024), with kind permission of Springer Nature, Cham

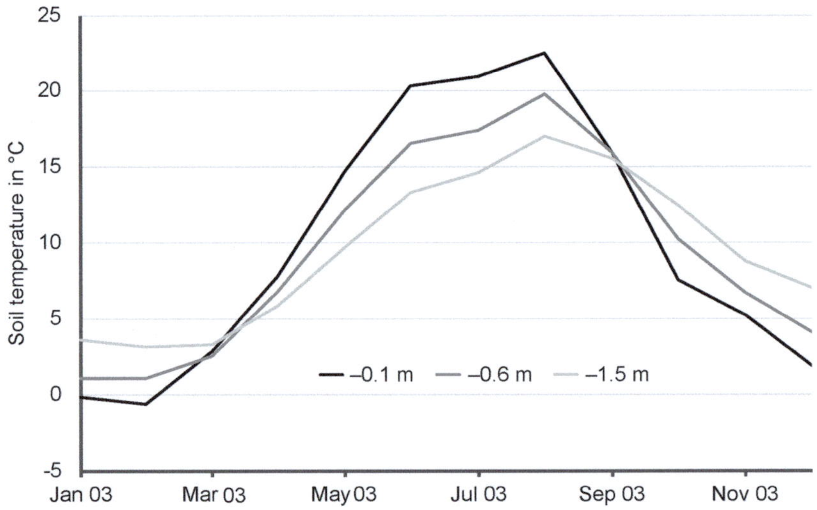

Fig. 1.10 Monthly mean ground temperatures under meadow near the Boundary Layer Measurement Field of the Meteorological Observatory Lindenberg, German Meteorological Service use of a data set from Beyrich and Adam (2007), according to Foken and Mauder (2024), with kind permission of Springer Nature, Cham

The following rules-of-thumb can be drawn: The drier a soil surface layer is, the less energy it can store and, more importantly, little energy can be transferred to deeper soil layers due to the reduced water and increased air content. Thus, the soil surface heats and more strongly, which leads to a large sensible heat flux and thus to heating of the air. Below the surface layer, the soil is cool and usually more moist in case of previous rains. A typical example is the sand on a beach (or desert), the surface of which can easily reach temperatures of over 50 °C during a sunny day.

The moister a soil is, the more energy it can store and transfer to deeper layers. This means that the upper layer of soil heats up only slightly in this case, especially also because a lot of energy is needed to evaporate the water from the soil, thus cooling the surface.

1.1.4 Energy Balance at the Earth's Surface

In ▶ Sect. 1.1.1, it was shown that the radiation balance at the Earth's surface is composed of the shortwave and longwave radiation components. The Earth's surface receives more radiation energy than it emits, which results in a positive net radiation (see the ◘ Fig. 1.12). This energy gain for the Earth's surface is released back to the atmosphere mainly by the two turbulent energy fluxes (see ▶ Sect. 1.1.2), the sensible heat flux (air which is warmed up at the surface) and the latent heat flux (evaporation, evapotranspiration), or is conducted into the soil by the ground heat flux (see ▶ Sect. 1.1.3). This is described in the basic **energy balance equation** (see Note 1.7). According to the conditions or application scale and needs, the energy balance equation may contain further terms such as the amount of energy from the net radiation used by plants in the biomass accumulation (photosynthesis) process (mostly < 1% of shortwave radiation). Also, energy stored in the mass of structures above the ground, such as buildings, which is not counted for the ground heat flux or heat storage/source terms caused by other energy sources than solar radiation (e.g., geothermic sources, fossil energy burning by vehicles, or for heating in cities), can be additional terms (see also Note 1.8). Thus, the energy balance equation can be formulated as the law of conservation of energy for the Earth's surface.

Note 1.7 Energy balance equation:

$$-Q_s^* = Q_H + Q_E + Q_G + \Delta Q_s \quad \text{(N1.3)}$$

with net radiation ($-Q_s^*$), sensible heat flux (Q_H), the latent heat flux (Q_E, evapotranspiration), the ground heat flux (Q_G), and the total heat storage change in a unit of time of the ground surface layer (ΔQ_s).

◘ Figure 1.11 shows a typical diurnal cycle of the turbulent energy fluxes of sensible and latent heat together with the net radiation and the ground/soil heat flux. Thereby, the above-defined sign convention (see Note 1.5) is again clarified by the change of direction of the energy fluxes during the day and at night.

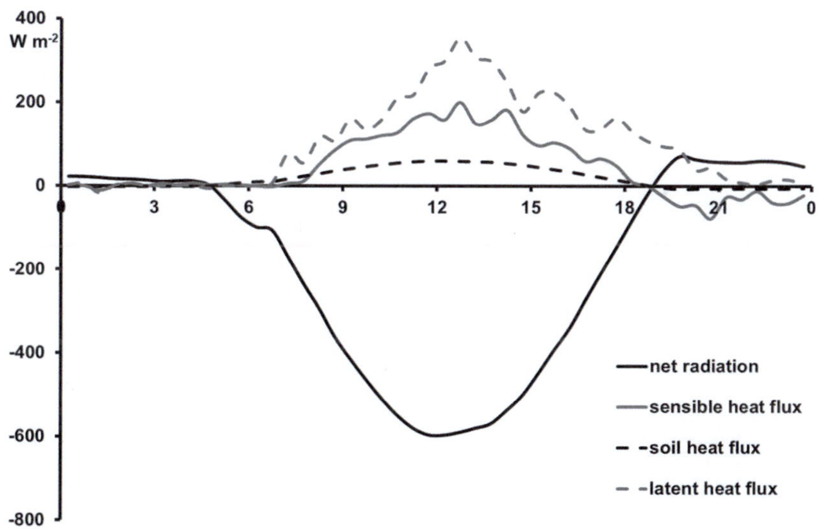

◘ **Fig. 1.11** Diurnal cycle of the components of the energy balance on 05.24.2012 in the Ecological-Botanical Garden of the University of Bayreuth (in CET, sign convention see Note 1.5), after Foken (2013) © Author, CC BY 4.0

1

The beginning of the air heating by the sensible heat flux starts already a few minutes after sunrise through heating of the energy exchange surface. The soil heat flux shows a longer reaction time. The turbulent energy fluxes usually reach their maximum after solar maximum on radiation (cloudless) days. While the net radiation shows a rather uniform diurnal cycle when the sky is cloudless, the turbulent fluxes are subject to stronger variations. The reason for this is the dependence of these fluxes on wind speed. Thus, in addition to sufficient water availability (soil moisture), air movement (wind) is also a prerequisite for evaporation.

A few hours before sunset, the energy from net radiation is often no longer sufficient to provide the necessary energy for potential evaporation. The surface then cools down more strongly, and the sensible heat flux already changes its sign, i.e., turbulence elements from higher, warmer air layers transport the necessary energy for the evaporation process in the direction of the surface (beside a potential contribution from the ground heat flux). This phenomenon, which is also typical for mid-latitudes, is called the **oasis effect** (see ▶ Sect. 1.2.7). As a result, the latent heat flux is usually still positive after sunset, while the sensible heat flux reaches its greatest negative values, and it changes its sign only in the early morning hours again with sunrise. Often, in the second half of the night, a short temperature increase in case of dewfall (condensation means release of energy) alternates in the immediate vicinity of the ground.

In deserts and arid zones, the sensible heat flux predominates. In mid-latitudes, the latent heat flux (Q_E) is usually about twice as large as the sensible one (Q_H), but also the surface conditions influence the ratio of sensible and latent heat flux, which is called **Bowen ratio** ($Bo = Q_H/Q_E$ or also known as "β"). Over the sea, evaporation predominates. Depending on the climatic region and weather conditions, the Bowen ratio can assume values of approximately 0.1–10.

On the global scale, the incoming shortwave radiation of the sun provides an annual average of 340 Wm^{-2} over the entire Earth (◘ Fig. 1.12). Not all of it reaches the Earth's surface. Some is reflected, some is absorbed by haze and clouds. The Earth's surface and also greenhouse gases, clouds, and haze emit longwave radiation according to their temperature. If one looks at the balance of all radiation fluxes at the Earth's surface,

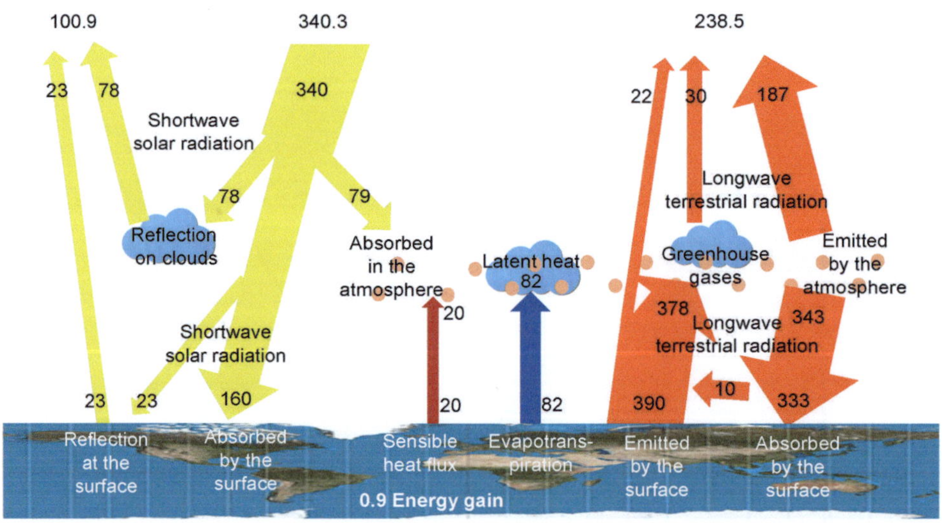

◘ **Fig. 1.12** Cycle of radiation and energy fluxes in the atmosphere (data in W m^{-2}) according to Trenberth (2022) based on the solar constant by Kopp and Lean (2011), graphic illustration according to Foken (2025).

the Earth's surface receives about 103 W m^{-2} more radiation energy than it emits. These 103 W m^{-2} are released back into the Earth's atmosphere by the turbulent fluxes of sensible and latent heat flux and a small quantity is stored mainly in the water of the oceans, thus completing the cycle (Fig. 1.12). This corresponds to approximately 30% of the total energy which comes from the sun (340 Wm^{-2}), supporting the importance of these micrometeorological surface processes. The partitioning of this 103 W m^{-2} into sensible (convective) heat flux and latent heat flux (evapotranspiration) varies greatly depending on the climate zone.

The additional anthropogenic greenhouse gases affect this cycle primarily through an increase in downward longwave radiation. This directly effects (as an additional energy source) the turbulent fluxes, with the fraction in the sensible heat flux expressing itself as global warming and increasing in the global mean temperature, and thus the term Q_H in the Bowen ratio (Wild 2020). Indirect effects also need to be taken in account for the impact on the latent heat flux potential (Note 1.9). In this context the soil heat flux is negligible in the annual mean (▶ Sect. 1.1.3).

Note 1.8 Energy balance:
When measuring all components of the energy balance equation above the Earth surface, **the energy balance may not be closed**. This is due to small-scale circulation systems whose energy transport is not captured by the measurements near the surface (Mauder et al. 2020).

Note 1.9 Clausius-Clapeyron's equation:
The equation describes the water vapor absorption of the atmosphere as a function of the air temperature. It means that with a linear air temperature increase, the capacity of the air to contain water vapor increases exponentially and thus also the potential capacity of water to evaporate. Thus, the temperature increase leads to higher (and faster) evaporation and increasing droughts also under unchanged precipitation pattern.

1.1.5 Water Balance at the Earth's Surface

As already mentioned, the radiation balance at the Earth's surface determines the energy available for the energy balance components, where the evapotranspiration (latent heat flux) term is the direct link to the water balance (Note 1.10, Fig. 1.13a). Precipitation provides the natural water supply from the atmosphere for plants and soils, at times supplemented with irrigation. Evapotranspiration and surface runoff as well as changes in the amount of water over time in the soil column are the other main parameters to be considered in the water balance of a certain spatial unit. The water balance equation is applied (and modified) for different spatial and time scales as well, such as in agrometeorology and hydrology (i.e., from plots or crop fields up to whole river catchments or regions).

Note 1.10 Water balance equation:

$$0 = P - Q_E - A \pm \Delta S_w \qquad \text{(N1.4)}$$

with the precipitation (P), the runoff (A), the evaporation (Q_E), and the water storage change primarily in soil and groundwater (ΔS_w).

If we analyze the energy budget of the surface, evaporation appears as latent heat, i.e., as energy transport. Therefore, it is possible to study how much energy is used for water evaporation in 1 day as well as to estimate the related water amount, which is more relevant for analysis of the water cycle and the water balance components (Note 1.11).

1

> **Note 1.11** Evaporation equivalent:
> The difference in terminology and units of measurement between those involved in micrometeorology and those involved in hydrology and agriculture should be kept in mind:
> - The water balance terms can be given in amount of water (e.g. mm; 1 mm = 1 L m^{-2}) or as amount of energy (e.g. Watt or Joule, where 1 W = 1 J s^{-1}) required for evaporation (1 W m^{-2} = 0.0347 mm d^{-1}).
> - The heat of evaporation (L, for latent heat) gives the energy required to evaporate 1 kg (=1 L) of water.
> - L depends on the water temperature and equals 2.45×10^6 J kg^{-1} (or J mm^{-1}) at 20 °C.

Considering the water balance components in the soil-crop system in detail, however, becomes more complex (◘ Fig. 1.13b). In the presence of plant canopies, precipitation reaches the ground only partially, because it is partly retained by plant surfaces (interception) and evaporates from these surfaces again. Like soil surface water runoff, this part of water will therefore not infiltrate into the soil. The amount of intercepted wa-

a)
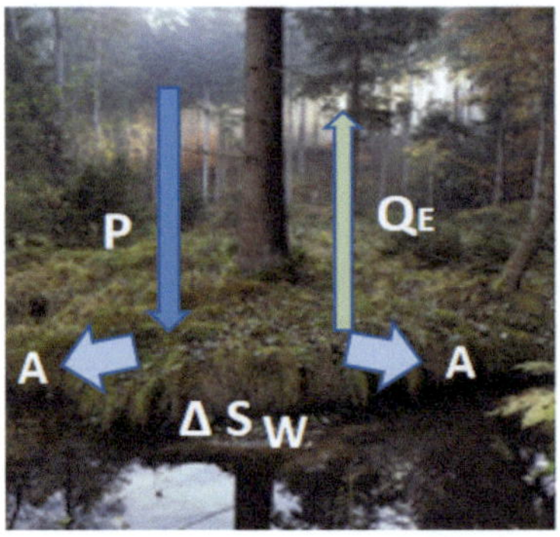

b)
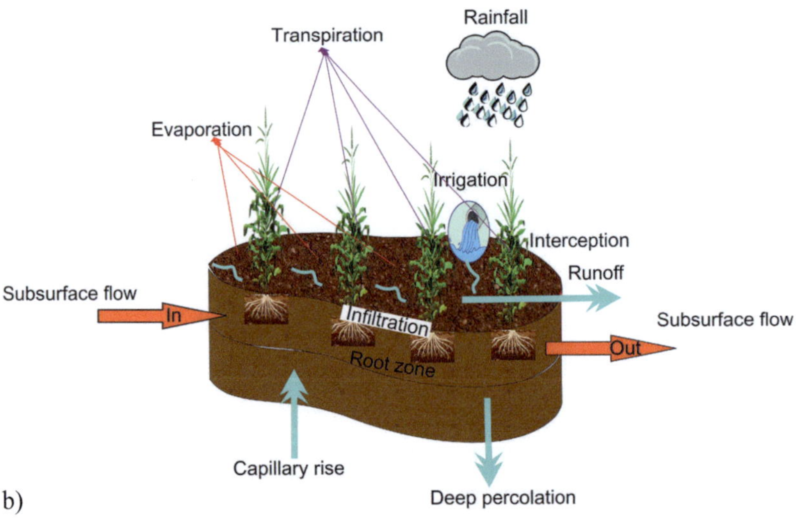

◘ **Fig. 1.13** **a** Basic water balance components (P is precipitation, A is surface runoff, ΔS_W is soil water storage change in a certain time period, and Q_E is evaporation from soil and other surfaces as well as plant/vegetation transpiration) (Photo: Eitzinger); **b** Detailed broken-down water balance components in the soil-crop environment (Lalic et al. (2018)

ter depends largely on plant stage, leaf area, canopy structure, and precipitation characteristics. On average, fully developed crop canopies reach interception levels of about 20%—in forests up to 80%—of total precipitation under Central European conditions.

When water from precipitation reaches the soil surface, infiltration into the soil and runoff depends on soil properties (physical, biological, and chemical soil characteristics). For soil water infiltration, pore size distribution at the soil surface as well as the soil surface structure play crucial roles, including aspects such as biotic pores (especially from rainworms), humus content, and the stability of pores during rains. Furthermore, soil gaps, caused mostly by drying clay minerals, can significantly change the amount of water which can infiltrate the soil over a certain period of time (infiltration rate).

Within the soil, the water movement is driven by gravity as well as by the capillary forces and the matrix water potential, which are determined by the physical (such as soil pore volume distribution) and chemical (e.g., salt content) soil properties, which can vary in the space significantly. The total soil pore volume determines how much water the soil can take up in total. However, only a certain amount of water is retained by the soil as capillary and adsorption water, which is partly available for plants (see ▶ Sect. 2.3.1). The plant's available water also depends on the rooting depth (and volume) of the specific plant/crop, where also additional components of the water balance of a soil column such as subsurface flows, which are often present in hilly terrains, can play a role.

Out of a certain soil volume (e.g., rooting zone), water can drain or percolate into deeper soil layers and reach the groundwater level, or water can rise up by capillary forces into the rooting zone from wetter soil layers or from the groundwater level.

As the determining soil properties for soil water balance can be highly variable in space, the installations of related soil water measurements in the soil column have to be planned carefully (exact placement of sensors, number of replications needed, and others; see Sects. 3.9 and 3.10).

1.2 Conditions for Measurements in the Atmosphere

The vertical and horizontal structure of the atmosphere and the typical time scales of atmospheric phenomena are important for the design and performance of measurements in the atmosphere, and especially in the boundary layer near the ground. This has an essential influence on the required data quality and on their usability for different purposes. For instance, the extension of different atmospheric layers and vertical gradients of thermodynamic variables have to be taken in account when deciding on the measurement range, measurement levels, and the vertical spacing of profile measurements. The time scales of processes determine the measurement frequency, averaging times, and response time of sensors.

The uncertainty of a meteorological measurement is not only the possible error of the sensor, but it is also dependent on the variability of the meteorological elements in space and time. Some typical situations and examples together with its theoretical basics are discussed in the following sections.

In order to carry out micrometeorological measurements in the atmosphere, ▶ Chap. 2 provides more information on the parameters and processes to be taken in account for measurements above various surfaces (see ◘ Table 1.3).

1.2.1 Structure of the Atmosphere

Atmospheric measurements in this book are mainly related to the troposphere, which exhibits a thickness between 5 km (polar regions) and 16 km (tropics). Most of the weather phenomena occur within this layer. The upper part of the troposphere is called free troposphere (free atmosphere) without the thermal and mechanical effects of the Earth's surface (without friction). It is mostly stably stratified, except in regions of strong deep convection. The wind in the free troposphere is basically determined from the pressure gradient and Coriolis forces (geostrophic wind).

Table 1.3 Atmospheric parameters and processes to be taken in account for micrometeorological measurements. The relevance for different underlying surfaces is given in Tables 2.1 and 2.2

Parameter	Section	Process	Section
Footprint	1.2.4	Stable stratification	1.2.6
Zero-plane displacement	1.2.3	Convection	2.2
Obstacles	1.2.5	Coherent structures	2.2
Internal boundary layers	1.2.5	Low-level jet	2.1
Heat sources	2.3	Afternoon transition	1.2.7
		Boundary layer height	1.2.1

The lowest part of the troposphere near the ground is called the **atmospheric boundary layer or mixed layer(ABL)**, on which these guidelines are focused. It is directly influenced by the Earth's surface and responds to surface forcing at time scales of about one hour or less (Stull 1988). The ABL has a typical thickness of about 1–2 km during daytime over land and of about 0.5 km over the oceans. In the daytime ABL, the potential temperature and the specific humidity profiles are nearly constant with height due to strong vertical mixing of air by turbulence. It is usually capped by a statically stable layer or temperature inversion. Strong gradients of temperature, humidity, and wind often occur across a relatively shallow interfacial layer separating the ABL from the free troposphere. This layer is part of the entrainment zone often combined with a capping inversion, where air from the free troposphere is mixed into the ABL (Deardorff 1966). The thickness of the entrainment zone typically is approximately 10% of the ABL depth, but it may be much larger under certain conditions. Entrainment processes have been proven even up to the surface layer. The night-time stable ABL over land has a typical depth of 100–400 m, but for strongly stable stratification, its thickness can be just 10 m or less especially in the polar regions. Turbulence is the most relevant process defining the structure and temporal evolution of the ABL. In the night-time the mechanical turbulence, in the daytime the thermal and mechanical turbulence setting up the exchange processes in the near surface soil-biosphere-atmosphere system.

The lowest part of the ABL is called the **surface layer** (SL); it roughly covers the lowest 10% of the ABL. Historically, this layer has been called Prandtl layer or constant-flux layer because the turbulent fluxes were assumed to be approximately constant with height within the SL, which is, however, valid only in the case of flat surface. This offers the possibility to estimate surface turbulent fluxes from vertical gradient or profile measurements of mean variables within this layer. The SL is turbulent to a large degree, and only within a few millimeters above the surface molecular exchange processes dominate. Above tall vegetation or urban areas, the constant flux layer assumption for flat surface is not valid due to the high friction, and the surface layer must be further subdivided (see Fig. 1.14). Immediately above the canopy up to roughly twice of the canopy height, in the roughness sublayer (Garratt 1978), also called **mixing layer**, turbulent mixing is increased. If the vegetation is not too high, a surface layer may develop above the mixing layer, but its thickness is reduced.

1.2.2 Scales of the Atmosphere

Meteorological processes can be associated with typical spatial and temporal scales. The reason for this is the spectral organization of atmospheric turbulence and wavelike processes, where relevant wavelengths (spatial extensions) are related to distinct durations in time (frequencies).

The principle of scale classification was formulated by Orlanski (1975). A close link

Magnitude of height in m	Layer of the troposphere	
10,000	Free troposphere	
1,000	Capping inversion	
	Upper layer	
10–100	Turbulent layer	Surface layer
	Roughness sublayer (mixing layer)	
0.01	Viscous sublayer	
0.001	Laminar boundary layer	

Fig. 1.14 Structure of the troposphere and of the ABL, following Foken and Mauder (2024), modified. The grey area is the relatively thin capping inversion (interfacial layer) between the free troposphere and the upper layer. The roughness sublayer is not developed over short vegetation

typically exists for atmospheric processes between their characteristic spatial and temporal scales, i.e., large-scale spatial phenomena typically have a long duration, and local processes occur at short time scales. The smallest range represents the processes in the boundary layer (this is what micrometeorology is concerned with), followed by mesoscale processes from small thunderstorms to tropical cyclones.

This is different for other compartments of the Earth system. For instance, hydrological processes in soils and plant ecosystems may represent large time scales, but at rather small spatial scales (Fig. 1.15). Therefore, it is an experimental challenge to combine measurements in the soil-vegetation system with measurements in the atmosphere. Examples of relevant scale dependent processes are presented in Table 1.4. Some processes may occur at different scales depending on the specific situation, e.g., the extent of a sea-breeze circulation depends on the size of the water and land bodies, on the thermal differences, and on the background wind regime.

1.2.3 Influence of Vegetation

Figure 1.15 is based on the assumption that the canopy has no vertical extension. However, low and dense vegetation can be considered as a porous medium, and the stand height (z_B) can be taken in account in the equations for momentum, heat, and mass transfer by an imaginary zero level. In case of sparse, inhomogeneous crop vegetation the zero level must be shifted upwards by the so-called **displacement height *d* (zero-place displacement)**. Thus, two height scales are defined, the geometric one with the zero level directly at the soil's surface and the aerodynamic one with the zero level at the height of the displacement height. Figure 1.16 illustrates this. However, it is not easy to specify the displacement height exactly. An approximate assumption is that the displacement height is two-thirds of the existing canopy height.

This method, however, is not possible to be used directly for high vegetation such as forests, tree plantations, and urban areas. These have a very high roughness, so that the airflow generates eddies. This is called the rough sublayer, which reaches a height of about 3 times the height of the structures, such as trees or houses. This produces an effect similar to that of unstable stratification in that turbulent exchange increases while vertical gradients decrease. From a hydrodynamic point of view, the stronger turbulence in this layer can be explained by an instability of the flow due to the strong vertical

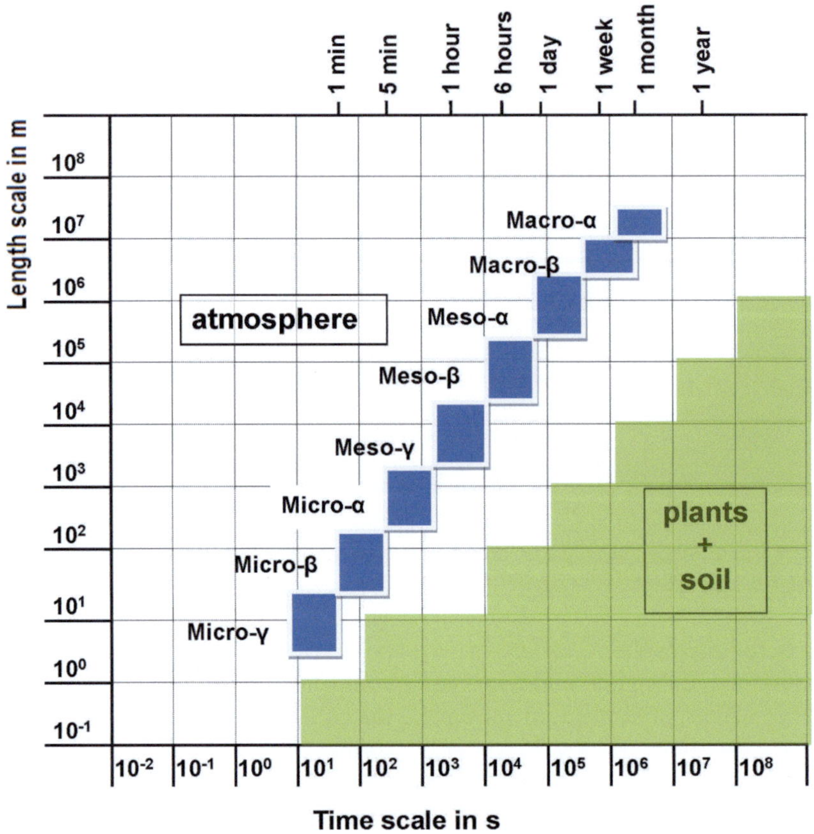

Fig. 1.15 Temporal and spatial scales of atmospheric, biospheric, and soil processes. Atmospheric scales (Orlanski 1975) are blue boxes with a size of one decimal magnitude (from micro-γ to macro-α) shown. The scales of the plant and soil processes, relevant for the energy and matter exchange with the atmosphere, are shown as green area after Foken et al. (2012, modified), © Author(s), CC Attribution 3.0 License

Table 1.4 Relevant scales in micrometeorology of atmospheric processes and phenomena, following Orlanski (1975), updated

Scale	Time	Space	Process
Meso-γ	30 … 180 min	2 … 20 km	Thunderstorms, deep convection, urban effects, clear air turbulence, thermally driven flows (valley winds, sea breezes)
Micro-α	5 … 30 min	200 … 2,000 m	Gravity waves, tornados
Micro-β	1 … 5 min	20 … 200 m	Wakes, coherent structures, dust devils, boundary-layer turbulence
Micro-γ	<1 min	<20 m	Roughness, surface-layer turbulence

gradient of the wind speed, so that it is also called a mixing layer (Raupach et al. 1996); see ► Sect. 2.1.2.

In addition, the canopy, but also a surface such as water or sand, is characterized by the so-called roughness height or roughness length (z_0, Table 1.5). The roughness length is a fictitious value if the logarithmic wind profile (see Note 1.12) with neutral stratification is extrapolated up to a height at

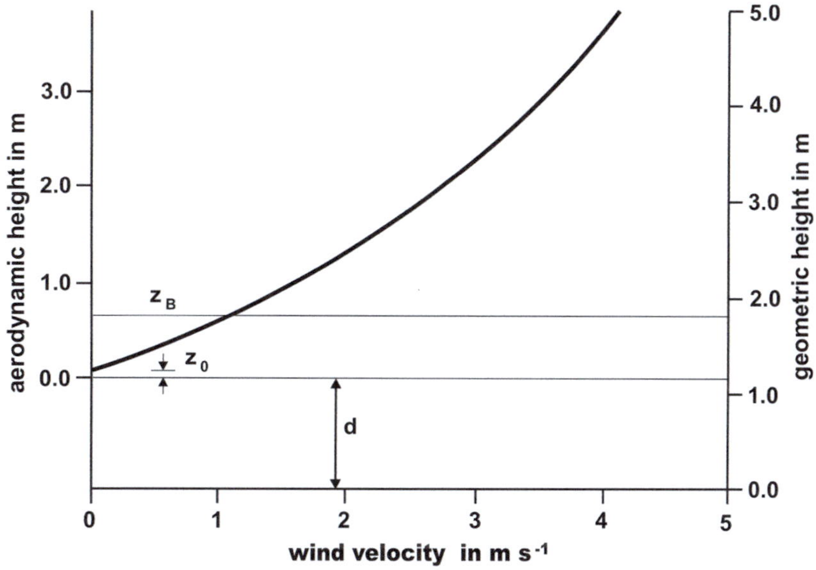

Fig. 1.16 The aerodynamic and geometric scale for a plant stand with, e.g., a canopy height z_B = 1.8 m and d= 1.2 m, from Foken and Mauder (2024), with kind permission of Springer, Cham

Table 1.5 Roughness height (length) in m for different surfaces (Foken and Mauder 2024, modified)

Surface	Roughness lengths
Ice	10^{-5}
Water	10^{-4}–10^{-3}
Snow	0.002
Grassland	0.005–0.02
Crops	0.05
Shrubs	0.2
Forest	1–2
Settlement	0.5–2

which the wind speed would be (but only theoretically) zero. This is a measure to determine the resistance that the surface exerts on the wind field, i.e., attenuates the wind close to the ground.

Note 1.12 The shape of the logarithmic wind speed profile with roughness length z_0 and displacement height d above the surface at height z is

$$u(z) = \frac{u_*}{\kappa}[\ln(z-d) - \ln z_0], \tag{N1.5}$$

where $\kappa = 0.4$ is the von Kármán constant, u_* is the friction velocity, which is proportional to the square root of the turbulent momentum flux.

1.2.4 Footprint

Measurements in the atmosphere differ significantly from those in solid media such as in the soil. Although atmospheric measurements above the ground take place at a very specific measurement location, the conditions at this location are often not known due to its potential high variation in time; instead, all the turbulence elements that pass the measurement sensor during a certain time are recorded within a specific measurement period. **This turns the point measurement into a spatial measurement.** As a rule of thumb, a measurement at a certain height is influenced by an area with an extension of about up to 100 times the measurement height in the direction of the wind. This area is smaller in case of unstable stratification and larger in case of stable stratification (and for high horizontal wind speed), and there are different influences from the roughness. In order to quantify these better, so-called footprint models are used. These models are based on models for the dispersion of air pollutants, whereby the calculation can be carried out using analytical equations or so-called **Lagrangian models**, in which particles are emitted in the model and their movements are tracked in the turbulent model atmosphere (Leclerc and Foken 2014).

The curve shown in ◘ Fig. 1.17 indicates the intensity with which certain areas have an influence on the measurement result at the actual location of the measuring device. Thus, the immediate nature of the ground near the equipment has little influence, while distant areas do contribute to the measurement result. Some footprint models are now available online on the Internet for simple applications (► https://footprint.kljun.net, or Spirig et al. 2017); see also (Chu et al. 2021). The footprint for scalars like temperature, humidity, etc. is much larger than for turbulent fluxes. When measuring turbulent fluxes, one has to be very careful of which surface the measurement is representative and of the respective location of the installation.

1.2.5 Internal Boundary Layers and Obstacles

The landscapes of the inhabited areas of the Earth are extraordinarily dissected. This means for the airflow that it has to adapt itself again and again to the surface. In the lower air layers, the airflow takes on the characteristics imposed on it by the roughness and thermal properties of the new surface. Above this, the turbulence elements still have the properties of the underlying layer in the wind direction. Between the air layer above the new surface (**new equilibrium layer**, Note 1.13) and the air layers above it, a disturbance layer is formed, which is called the **internal boundary layer.** In highly structured landscapes, multiple internal boundary layers can form (◘ Fig. 1.18), which mix at a certain height (blending height, about 50–100 m) in such a way that the turbulence elements take on the properties of all the surfaces. In situations of low wind and high solar radiation, however, turbulence elements may be exchanged only vertically. Then one can still detect the properties of the surface in the turbulence elements at heights of up to several kilometers. For practical purposes,

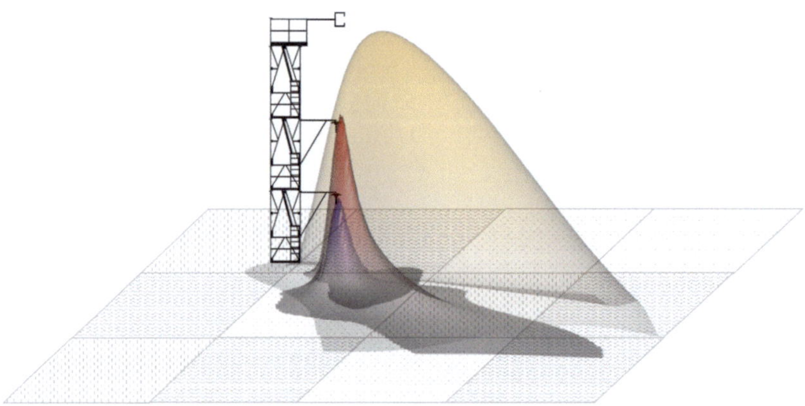

◘ **Fig. 1.17** Schematic representation of the footprint function for measurements in real landscape. After Metzger (2018) with permission from Elsevier

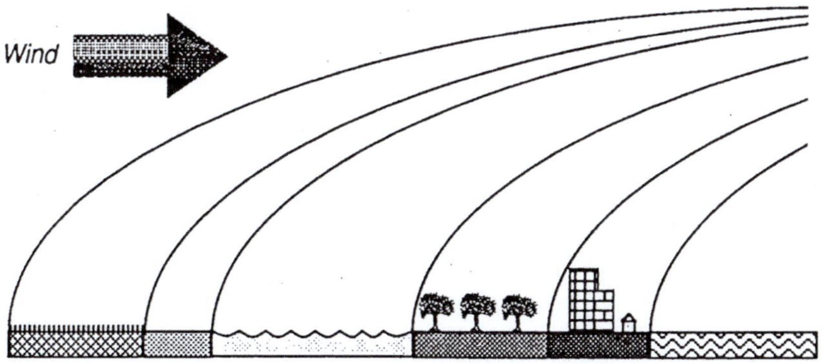

Fig. 1.18 Development of internal boundary layers in a heterogeneous landscape, from Stull (1988), with kind permission of Springer, Dordrecht

the height of the new equilibrium layer near the surface can be estimated in a simple way as a function of **the fetch** (distance of the measuring point from the last change of the substrate roughness) (e.g., the distance to the border of a homogeneous crop field) according to Raabe (1986).

Note 1.13 New equilibrium layer:
Simple approach to determine the height of the new equilibrium layer:

$$\delta = 0.3\sqrt{x} \qquad \text{(N1.6)}$$

where δ is the height of the new equilibrium layer and x the fetch.

If, for example, one wants to assign measurements exactly to a certain surface, one must not only consider the footprint, but must also determine the measurement height below the height of the new equilibrium layer.

At a certain height (about 50–100 m), above the internal boundary layers, the so-called **blending height** (Mason 1988) measurements are assumed to be representative of a larger area than just represented by the local underlying surface. This is relevant for profile measurements at tall towers or using **ground-based remote sensing techniques**, as well as for measurements along horizontal paths (so-called transects) with aircraft/drones or scintillometers (see ► Sect. 3.12).

In the case of single obstacles in the landscape (buildings), one can assume that the turbulent flow field is already disturbed at 2 times the obstacle height before the obstacle and 20 times the obstacle height after the obstacle, and that the disturbance is still pronounced at 2 times the obstacle height. Flux measurements should not be carried out in such disturbed areas; for the measurement of turbulent fluxes, the disturbed areas must be set at a factor of 5–50 times the obstacle height.

1.2.6 Gradients Near the Surface

Figures 1.19 and 1.20 show typical vertical wind and temperature profiles near the surface for light wind conditions (for nearly calm conditions at night, the gradients under stable stratification are significantly larger) for stable, neutral, and unstable stratification and typical differences of the wind speed and temperature at different levels in comparison to the meteorological standard measuring levels at 10 m for wind and 2 m for temperature (WMO 2024). From these profiles, some implications for the performance of atmospheric measurements can be derived. For example, for neutral stratification over grass, the wind speed at 2 m height is only 77% of the wind speed at the reference height of 10 m. The typical measuring error for temperature of 0.1 K allows a tolerance for the standard measurement height (2 m) of about ±0.2 m.

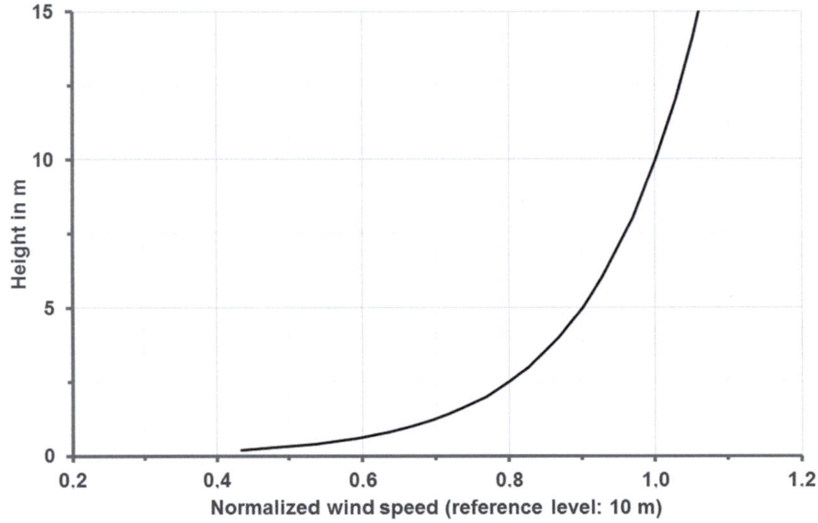

Fig. 1.19 Wind speed profile normalized to the value at 10 m for neutral stratification over short grass ($z_0 = 1$ cm); for slightly stable and unstable stratification the profiles are similar; from Foken et al. (2021b), with kind permission of Springer, Cham

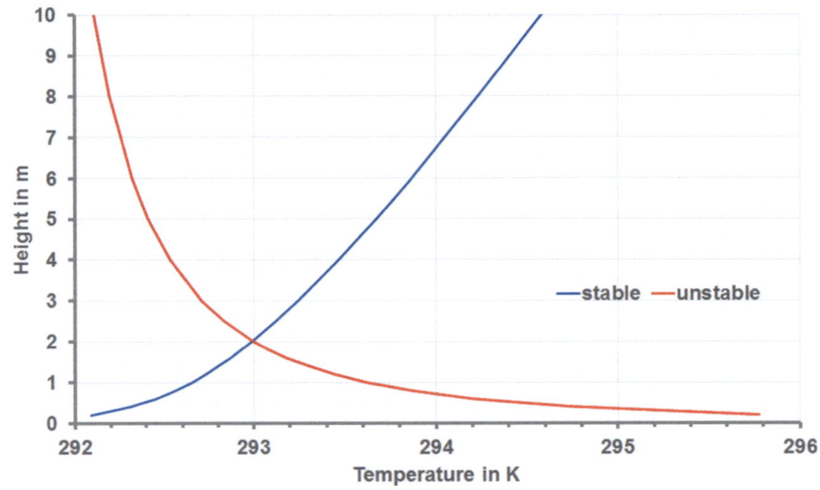

Fig. 1.20 Typical temperature profiles for stable and unstable stratification with light wind speed; from Foken et al. (2021b), with kind permission of Springer, Cham

1.2.7 Afternoon Transition and Oasis Effect

During micrometeorological measurements, typical cases will always occur that are not immediately recognized. While the sensible and latent heat flow increase in the morning with the onset of sunlight, the sensible heat flux becomes negative before sunset. However, the latent heat flux remains positive for several hours after sunset (see Fig. 1.11). The energy required for evaporation is provided in this case by the downward sensible heat flux and the heat flux from the ground. This phenomenon is known as **the afternoon transition**; and the downward sensible heat flux (i.e., energy transport) and the use of this energy for evaporation is also known as **the oasis effect** (Stull 1988). The term is derived from the conditions in an oasis when dry and hot air stream (from the desert) over a moist, evaporating surface. A similar effect can also occur at irrigated fields surrounded by dryer and warmer areas.

1.2.8 Horizontal Variability of Meteorological Elements

Measurements at a given site can be considered as spatially representative if the measured variable does not vary significantly over a certain distance around the site (Foken et al. 2021b). The typical spatial variability differs significantly for different elements, and it may also depend on the meteorological and site conditions. For instance, global and diffuse radiation show very little spatial variability in the absence of clouds and over flat terrain, but larger differences occur in mountainous regions, and very large differences can be observed under conditions of variable cloud cover and, of course, below plant canopies. Net radiation is strongly dependent on the albedo of the underlying surface and on its temperature and wetness. It may show considerable spatial variability even if the downward short- and longwave radiative fluxes appear to be horizontally homogeneous.

The wind field is basically determined by the large-scale pressure distribution, but it is modified close to the surface by the topography, surface roughness, and obstacles (e.g., trees), which can lead to significant small-scale differences.

Air temperature and relative air humidity temporal variability are closely related to both the changes of net radiation and air mass characteristic (absolute water content of the air). In a uniform air mass and at the same height, these show very little horizontal differences, except close to the surface, in particular during high insolation and during nights with significant radiative cooling. Thus, the nocturnal temperature minima (and related relative air humidity) near ground may show remarkable differences even over small distances during the diurnal cycle.

Considering the temporal scale (wind related) frontal drifts, thunderstorm activity, or precipitation cooling, the surface can lead to more rapid changes in temperature and relative humidity.

Horizontal variability of precipitation strongly depends on the synoptic situation. Steady rain ahead of a warm front often results in a relatively homogeneous horizontal distribution of the measured precipitation amount while very strong small-scale differences can occur in connection with showers or thunderstorms. Precipitation patterns are also strongly influenced by orography and land surface heterogeneity. Even human activities, such as irrigation in agriculture, can rise such strong small-scale differences, often spread over larger irrigated regions.

Major reasons for a possibly significant spatial variability of meteorological elements close to the surface are summarized in ◘ Table 1.6.

1.3 Keeping Standards and Rules in Measurements

For mutual understanding and communication of humanity, it is indispensable to establish standards and rules, such as a common language or traffic rules. The same principle is true in technics and science, where we need common protocols, standards, or rules to be able to transfer information, data, and knowledge which can be understood, accepted, used, and trusted by a wide audience.

It means that the way how we measure environmental phenomena should keep accepted standards of different types. Such standards for (micro)meteorological technics, sensors, and methods were developed at the global level (e.g., by World Meteorological Organization—WMO) as well as for specific applications (such as measurement networks, measurement tasks) and specified as—depending on the application—strict protocols, minimum requirements, or just recommendations (Jäckel et al. 2021).

1

Table 1.6 Aspects of micro- and meso-γ scale horizontal variability of meteorological elements near to the surface outside of forests and urban areas (Foken and Mauder 2024)

Meteorological element	Range and reason of variability	Conditions for low variability
Global and diffuse radiation	Generally small	Free horizon, no clouds
Net radiation	Partly significant due to differences in albedo and surface temperature, in mountain regions extremely	Unlimited horizon and uniform underlying surface
Wind velocity and wind direction	Partly significant in complex terrain, over strongly varying land-cover type, roughness, and in the case of obstacles	Large fetch over uniform surfaces, no obstacles
Temperature and air humidity	Often small	Open, flat landscape, except local convective precipitation or human intervention (irrigation)
Minimum temperature, temperature near the surface, temperature of the upper soil layers	Partly significant, especially in valleys and hollows (also at very small scales), mainly for stable stratification	Open, flat landscape, except local convective precipitation or human intervention (irrigation)
Precipitation	Partly significant, especially under convective conditions in summer, otherwise mostly in the range of the measurement error	Open, flat landscape, except local convective precipitation or human intervention (irrigation)

In planning measurements, we first need to deal with the scale-dependent horizontal variability of meteorological elements and the representativeness of measurements for the specific application or task. Based on this, we can define the needed type and characteristics of sensors and the selection of measurement sites. Additionally we have to design the measurement program in order to fulfil the planned task, considering, for example, the required number of measurement sites and the duration of the measurements. This will further lead to consider economical questions (see ► Sect. 1.3.4) related to investment **and the often underestimated maintenance and personnel costs for quality assurance**. The overall aim is to optimize economic efficiency while maintaining the expected results in quality and quantity.

It is not enough just to purchase the measuring instruments and install the stations. This is merely the first step. The work is qualified by the results of the measurement program (quality-assured databases, satisfaction of user needs, actual application in decision/production processes, and, for scientific programs, peer-reviewed publications). The measurement system must/should be designed for the whole foreseen lifetime of the measurement system.

1.3.1 Representativeness of Meteorological Measurements

In order to obtain qualitatively good and representative data, this requires that, on the basis of the measuring task, the locations, and the measuring principle, the measuring instrument properties and the structure and dynamics of the meteorological elements and fields must be harmonized with each other in a suitable manner (Foken et al. 2021b). From this, a suitable measurement execution including coordinated evaluation procedures must be derived.

The choice of the measuring systems and the location is affected by three basic questions which can be answered in different ways, so that an optimal and problem-adapted solution must be found between them.

- How do the horizontal and vertical variations of meteorological fields affect placement? The structure of the meteorological fields and the individual meteorological elements such as radiation, temperature, humidity, and wind should be assessed beforehand approximately with regard to their horizontal differences and vertical gradients in order to position measuring instruments in such a way that the desired information about the distributions of the variables can also be measured. This might need the help of experts, in case of missing own know-how.
- Should the measurements represent a larger area or specific local effects? Requirements for the measurements are determined by the question of whether they should provide information for a larger area or whether small-scale, local effects should be considered. This is also connected with the question for which area or vertical layer a measurement should be representative.
- What is the optimum compromise between using high-accuracy systems and sensors versus cost-effective ones?

For example, depending on the problem to be investigated and its relevance, measurement tasks can be carried out using standardized, quality-assured individual systems or with a large number of distributed sensor systems that meet only limited accuracy requirements but are cost-effective (see ▶ Sect. 1.3.4).

The guidelines of the World Meteorological Organization (WMO 2024) specify requirements for measurements that serve the purpose of long-term climate monitoring and meteorological forecasts with internationally standardized instrument types and installation conditions. For many tasks in applied meteorology and research, these guidelines can be used as an orientation, but must be adapted to the object of investigation in individual cases. The spatial and temporal representativeness of any atmospheric measurement performed using in situ or remote sensing measurement techniques has to be considered. Spatial or local representativeness is important in all applications (Foken et al. 2021b).

The representativeness of a site for a measurement (see further details in ▶ Sect. 2.7) can be determined according to two aspects:

- Representativeness for a larger area, as is necessary especially for measuring stations for synoptic and climate measurements. Often a distinction is made between rural and urban locations.
- Representativeness for a specific question, application, or location, where other influencing factors are to be largely excluded.

The **measurement height is very important** for the representativeness of measurements to ensure comparability, especially for temperature, air humidity, and wind. The measuring height for these variables is standardized for climatological weather stations of national weather station networks worldwide, which is 2 m above ground (short grass) (1.25–2.0 m, WMO 2024) and for wind also 10 m above ground, so that vertical gradients hardly distort the measurements (see ▶ Sect. 1.2.6). Other defined measurement heights are applied for precipitation, evaporation (e.g., Class A Pan), or near-soil surface temperatures. Additional, different measuring heights can of course be selected, depending on the specific purpose at standardized weather stations. Besides the measurement height of standardized climatological stations other requirements, such as the local position and distance to disturbing obstacles, the surrounding area conditions, minimum maintenance of the measurement site, technical aspects such as regular sensor calibrations and sensor technology, and precision have to be fulfilled.

In general, measuring instruments are fixed mounted at a certain height above ground (geometric height, see ▶ Sect. 1.2.3). This is not a problem for a surface where the displacement height doesn't change over time (e.g., due to growing plants) such at WMO standard weather stations which have to keep growing grass always short. However, for many applications the height above the displacement height has to be the reference measurement height (e.g., when models require or are calibrated for using wind in 2 m height for calculation of evapotranspiration and others). Since the displacement height can change at sites out of the WMO standard with the long-term (e.g., trees, forest) or seasonal growth (e.g., annual crops) of a plant stand, the measuring height also changes. In this case, the sensor height must be shifted in height accordingly, e.g., by applying a lifting mast (Kolle et al. 2021), if the specific application or research question requires it.

Besides the global climatological station networks, **different applications or problems** can define however their own "standards" (see ► Sect. 2.7), depending on the specific demands or research questions.

Practical applications in **rural areas** often are related to agronomic problems. For example, there is the need for temperature, air humidity, or leaf wetness measurements within the microclimate of crop canopies for plant protection methods/models or the measurement of soil-crop water balance variables within a crop field (soil water content and tension, evapotranspiration, etc.) for irrigation planning. Due to the manifold dynamics and structures of crop/plant/vegetation canopies, these applications have to apply different rules for appropriate placement of sensors and measurement height. For example, soil water sensors for irrigation planning are applied within the rooting zone depth of the specific crop, precipitation and global radiation sensors for calculation of evapotranspiration have to be placed above any maximum canopy height, and air humidity sensors (or others) for pest protection are placed in that part of the canopy, where the pests also occur.

If measured data are used as (agronomic) model inputs, the expected measurement standard of input data need to be always considered, however. On the other hand, if measured data will be used for model validations, the measurement height, or soil measurement depth, should fit to the model outputs.

Measurements for applications in **urban areas** are of particular interest. Here, a target must be clearly formulated through quality management and the measurement must be carried out in such a way that it fully serves the purpose, and other influencing factors are largely eliminated. Increasingly, "low cost" sensors are widely available or are built into devices such as mobile phones and cars. If measurement data are available in bulk, mathematical models are used to try to generate usable measurement results. This is called crowdsourcing (Muller et al. 2015; Budde 2021).

Not only for rural and urban applications but in general for all kinds of meteorological measurements, it is important that meta-data and meta-information (see Note 1.14) on equipment and location conditions are available in order to be able to assess the measurements in terms of their usability and quality. Furthermore, reliable reference stations should be available.

Note 1.14 Meta-data:
In general, it has to be kept in mind that meta-data/information of any measured data set is a crucial part of data quality assurance. For example, the use of measured data sets for any application or a comparison of different measured data sets is not possible if the circumstances of the applied measurement methods, the station design, and the measurement sites are un-known. It's imperative for a person which installs/plans measurements to make sure to record and write down as much as possible details of meta-data and meta-information from the BEGINNING of the measurement or station installation, ideally through a standardized protocol. Photo documentation of the installations, site conditions, and other relevant aspects are among others strongly recommended (see more details in ► Sect. 4.4.2).

1.3.2 Selection of Suitable Measurement Sensors

Before selecting suitable measurement sensors, it is necessary to be clear about which temporal and spatial differences of a meteorological variable are to be measured in order to unambiguously record a process or to be able to provide input data for calculations and modeling.

Table 1.7 gives an overview of technical characteristics of common measuring instruments of the most important meteorological variables. Besides the technical uncertainty, the response time (time constant, Note 1.15) of sensors, from which the temporal resolution results should be considered for the selection. The sampling rate of the data logger must in any case be significantly higher (shorter time differences) than the time constant of a sensor. For turbulence measurements, an even more careful adjustment must be made (see ► Sect. 3.12). Siting recommendations and measurement ranges of different sensor types are given in Table 1.8.

Note 1.15 Time constant:
Time after the expiration of which the measurement signal indicates $(1-1/e)\times 100$ (which corresponds to 63%) of the total change after an abrupt change in the value:

$$X(t) = X_{\infty}\left(1 - e^{-\frac{t}{\tau}}\right) \tag{N1.7}$$

where $X(t)$ is the measured value at time t, X_{∞} is the final value, t is the time, and τ is the time constant.

In addition to the traditional stationary measuring stations, there are mobile measuring platforms such as mounted on cars, bicycles, and drones or even only carried by walking, which need to include ideally a GNNS system in order to be able to localize the measured data on maps and for interpretation (see example in ► Sect. 2.7.2).

Table 1.7 Selected important measuring instruments with their typical characteristics and specifications for standard measurements (Foken et al. 2021b)

Meteorological element	Sensor type	Uncertainty range	Response time (time constant)	Sampling rate (data logger)
Temperature	Pt-resistance with radiation screen	1 K	1 min	10 s
	Pt-resistance, ventilated with radiation screen	0.1 K	10 s	1 s
Relative humidity	Capacitive sensor, ventilated with radiation screen	1%	30 s	1 s
Wind speed and direction	Cup anemometer with wind vane	$0.2\ m\ s^{-1}$ 3°	10 s, velocity dependent	1 s
	Sonic anemometer	$0.1\ m\ s^{-1}$ 2°	<0.1 s	0.01 s
Pressure	Piezoelectric sensor	0.1 hPa	10 s	1 s
Precipitation	Tipping bucket rain gauge	0.1 mm, resolution about 10%	1 min	10 s
Global radiation	Pyranometer	$10\ W\ m^{-2}$	10 s	1 s

Table 1.8 Common sensors with siting condition and typical measurement ranges (VDI 2025)

Measurement variable	Types of sensor	Particularity	Measurement range	WMO (2024) recommended height for standard measurements
Air temperature	Resistance thermometer, thermocouple, bimetallic strip, liquid-in-glass thermometer	Mounted in a radiation shield with natural or forced ventilation	(–35 … 50) °C	2 m above ground
Air humidity	Capacitive sensor, hair hygrometer, psychrometer, dew-point hygrometer	Mounted in a radiation shield with natural or forced ventilation	(10 … 100) %	2 m above ground
Atmospheric pressure	Electrical sensor, aneroid barometer, mercury barometer (no longer permitted since 2021)	Can be used in ventilated indoor spaces	(800 … 1050) hPa	Pay attention to height dependence of atmospheric pressure
Wind speed	Cup anemometer, ultrasonic sensor	Make sure that there are no obstacles	(0 … 60) $m\ s^{-1}$	10 m above ground
Wind direction	Wind vane, ultrasonic sensor	Make sure that there are no obstacles	(0 … 360)°	10 m above ground
Amount/intensity of precipitation	Weighing-type precipitation gauge, tipping-bucket gauge, drop counter, collection container (amount only)	Make sure that there are no obstacles	up to 200 mm/(0…300) $mm\ h^{-1}$	1 m above ground
Shortwave radiation	Pyranometer (thermocouple), photodiodes	Make sure that the instrument has a free horizon and is ventilated, if possible	(0…1500) $W\ m^{-2}$	>2 m above ground
Longwave radiation	Pyrgeometer (thermocouple)	Make sure that the instrument has a free horizon and is ventilated, if possible	(200…500) $W\ m^{-2}$	>2 m above ground

1.3.3 Data Transmission Options and Relevance

Meteorological processes, weather phenomena, or any change in microclimatic variables happen all the time around the world. Sensors placed in the atmosphere (or below ground for processes which are also related to atmospheric events such as change in soil temperature and soil water content) measure various variables in certain time steps and deliver specific values. Using the nowadays widely distributed electronic sensor devices and stations, the digitalized data can be stored at the local station and be transmitted sooner or later to central data collection for further processing and use.

Specifically, electronic sensors produce analogue electronic output signals (continuous range of values to represent the data, such as a resistance or voltage) corresponding to the measured physical phenomenon. The electronic output of a sensor can be also an on–off switching (used, e.g., for tipping bucket rain sensors), what can be counted or its frequency can be calculated. To be able to store its values in digital numbers, the analogue value needs to be converted (using an analogue-digital, **A/D converter**, see ▶ Sect. 2.7.4) to a digital value. The A/D converter can be located in a data logger or—nowadays more and more common—in the sensor device itself, which means the sensor device cable output is already a digital signal (discrete 0 and 1, to represent the data) in a specified format corresponding to a protocol (◘ Fig. 1.21).

Usually, a special device called **data logger** (see ▶ Sect. 2.7.4) collects these (analogue as well as digital) signals, data from several various sensors, processes this data (minimum, maximum value, average, conversions), stores, and transmits forward to a remote location for further use.

The data collected and stored by a data logger of any station can be periodically “emptied”, “downloaded” by a human operator into a mobile storage (USB stick, RS232 or USB cable download to a Laptop, CF card, etc.), but it requires regular human intervention which can be costly and difficult in remote areas. Then the collected data cannot be processed or checked in real time (almost at the same time when the data from the sensor arrives) but only much later, days, weeks, or months after the collection.

This increases the risk for large data gaps, if sensors or full stations fail. However, the wired connection gives the opportunity to directly check measurements and download data for the on-site user. This can also avoid possible errors in online transfer connections. Devices that provide wired and wireless access in parallel are therefore recommended for safe and robust data collection.

The preferred method for the monitoring of data collection is the automatic data transmission, when the data is transmitted

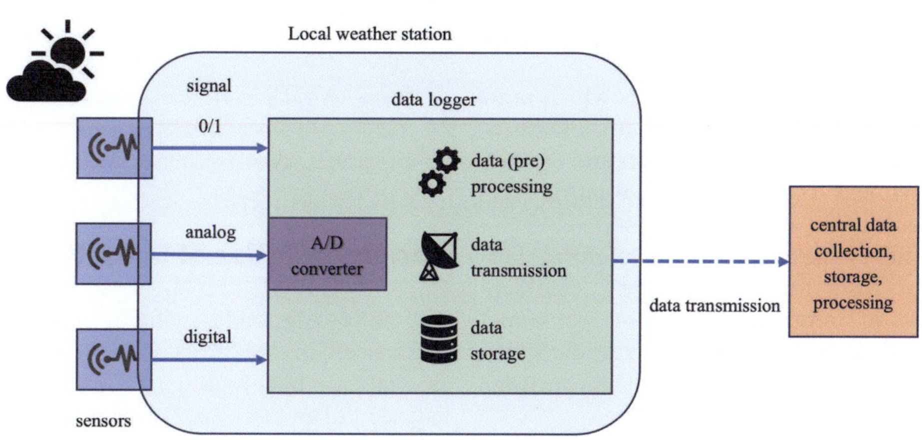

◘ **Fig. 1.21** Sensors, signals, data logger, and data transmission scheme. *Source* Zoltán Istenes

from the station (or also from a network of sensors) through various techniques without any human intervention, in real time or in packages (periodically from seconds to hours/days) to a central data location (internet servers) where data can be accessed for the users (for more detailed information, see ▶ Sect. 2.7.4).

1.3.4 The Economy of Measurements

The economy of measurements is strongly application-oriented, where for research applications or professional operational services often the highest possible or a mandatory standard of quality of measurements in respect of an available budget is the decision limit; for most practical applications in an economically driven sector, the cost-return of a certain measurable benefit from the measurements is the deciding factor. For the latter case, the total costs for measurements (or finally for the data which are used) play a critical role in order to make applications profitable, such as for applications in agriculture. A good example are station networks for plant protection services, which are proved to be one of the most economic applications in agriculture. Under these conditions, task-oriented measurement programs should be developed and established in the economic context, which could be one or more weather stations on a single site or local measurement networks managed by a group or organization, e.g., for rural or urban applications.

The three main types of costs which need to be taken into account for any measurement program and its application are

- Investment cost (costs of weather stations, sensors, installation costs).
- Maintenance costs for the planned lifetime (costs for hardware (e.g., replacement of sensors), costs for personnel (e.g., quality monitoring, repair, data curing), annual costs for online data transfer cloud services).
- Costs occurring for data application (e.g., personnel costs for data preparation and analysis, visualization, model applications, transfer of information to other users).

Often, the total costs of a measurement program can exceed the potential benefits. Therefore, it is important to calculate the economic balance carefully and consider low-cost alternatives before making an investment decision (Note 1.16). Beside the optimization of investment costs for sensors or weather stations, a partly use or replacement by alternative already available data sources is an option for cost reduction. These alternatives, however, should fulfil the required quality standard of the planned application at a relevant site or region. For a more detailed checklist on economic aspects to be considered, see ▶ Sect. 2.7.5; for examples of economic application, see ▶ Sect. 2.8.2).

Not only measured real-time data **could be provided from alternative sources** (see ▶ Sect. 4.8) **by lower costs**, there are also available already processed data for past (climatic) or future periods (climate change scenarios—long term, or meteorological forecasts—short term) which can provide information needed for many applications, such as for planning adaptation and mitigation options (e.g., in agricultural and urban management, environmental services) and any activity at different spatial and temporal scales along value chains for socioeconomic benefits.

> **Note 1.16** Data of poor or unknown quality is less useful than no data since it can lead to wrong results or decisions.

Appendix 1: Use of SI Units

The following table includes important SI units used in the book (SI 2019). The basic units are **bold** highlighted. For a complete list, see Foken et al. (2021a).

Name	SI Unit	Unit	Calculation
Length	**meter**	**m**	
Time	**second**	**s**	
Velocity		m s^{-1}	1 km h^{-1} = (1/3.6) m s^{-1}
Acceleration		m s^{-2}	
Mass	**kilogram**	**kg**	
Density		kg m^{-3}	
Impulse		kg m s^{-1}	1 kg m s^{-1} = 1 N s
Force	newton	N	1 N = 1 kg m s^{-2}
Pressure, friction	pascal	Pa	1 Pa = 1 N m^{-2} 1 Pa = 1 kg m^{-1} s^{-2}
Air pressure	hektopascal	hPa	1 hPa = 100 Pa
Work, energy	joule	J	1 J = 1 N m = 1 W s 1 J = 1 kg m^2 s^{-2}
Power	watt	W	1 W = 1 J s^{-1} = 1 N m s^{-1} 1 W = 1 kg m^2 s^{-3}
Energy flux density		W m^{-2}	1 W m^{-2} = 1 kg s^{-3}
Temperature	**kelvin**	**K**	
Celsius-temperature		°C	0 °C = 273.15 K
Temperature difference		K, (°C)	

References

Beyrich F, Adam W (2007) Site and data report for the Lindenberg reference site in CEOP: phase I. Ber Dt Wetterdienstes 230:55

Budde M (2021) Crowdsourcing. In: Foken T (Hrsg) Handbook of atmospheric measurements. Springer Nature, Cham, pp 1201–1233. ▶ https://doi.org/10.1007/978-3-030-52171-4_44

Chu H, Luo X, Ouyang Z, Chan WS, Dengel S, Biraud SC, Torn MS, Metzger S, Kumar J, Arain MA, Arkebauer TJ, Baldocchi D, Bernacchi C, Billesbach D, Black TA, Blanken PD, Bohrer G, Bracho R, Brown S, Brunsell NA, Chen J, Chen X, Clark K, Desai AR, Duman T, Durden D, Fares S, Forbrich I, Gamon JA, Gough CM, Griffis T, Helbig M, Hollinger D, Humphreys E, Ikawa H, Iwata H, Ju Y, Knowles JF, Knox SH, Kobayashi H, Kolb T, Law B, Lee X, Litvak M, Liu H, Munger JW, Noormets A, Novick K, Oberbauer SF, Oechel W, Oikawa P, Papuga SA, Pendall E, Prajapati P, Prueger J, Quinton WL, Richardson AD, Russell ES, Scott RL, Starr G, Staebler R, Stoy PC, Stuart-Haëntjens E, Sonnentag O, Sullivan RC, Suyker A, Ueyama M, Vargas R, Wood JD, Zona D (2021) Representativeness of eddy-covariance flux footprints for areas surrounding AmeriFlux sites. Agric Meteorol 301–302:108350. ▶ https://doi.org/10.1016/j.agrformet.2021.108350

Deardorff JW (1966) The counter-gradient heat flux in the lower atmosphere and in the laboratory. J Atmos Sci 23(5):503–506. ▶ https://doi.org/10.1175/1520-0469(1966)023%3c0503:tcghfi%3e2.0.co;2

Foken T, Meixner FX, Falge E, Zetzsch C, Serafimovich A, Bargsten A, Behrendt T, Biermann T, Breuninger C, Dix S, Gerken T, Hunner M, Lehmann-Pape L, Hens K, Jocher G, Kesselmeier J, Lüers J, Mayer JC, Moravek A, Plake D, Riederer M, Rütz F, Scheibe M, Siebicke L, Sörgel M, Staudt K, Trebs I, Tsokankunku A, Welling M, Wolff V, Zhu Z (2012) Coupling processes and exchange of energy and reactive and non-reactive trace gases at a forest site—Results of the EGER experiment. Atmos Chem Phys 12(4):1923–1950. ▶ https://doi.org/10.5194/acp-12-1923-2012

Foken T (2013) Energieaustausch an der Erdoberfläche. Edition am Gutenbergplatz, Leipzig. ▶ https://doi.org/10.15495/EPub_UBT_00006722

Foken T, Hellmuth O, Huwe B, Sonntag D (2021a) Physical quantities. In: Foken T (Ed) Springer handbook of atmospheric measurements. Springer Nature, Cham, pp 107–151. ▶ https://doi.org/10.1007/978-3-030-52171-4_5

1

Foken T, Beyrich F, Wulfmeyer V (2021b) Introduction to atmospheric measurements. In: Foken T (Ed) Springer handbook of atmospheric measurements. Springer Cham, pp 3–31. ▶ https://doi.org/10.1007/978-3-030-52171-4_1

Foken T, Mauder M (2024) Micrometeorology, 3rd edn. Springer, Cham. ▶ https://doi.org/10.1007/978-3-031-47526-9

Foken T (2025) Bamberg im Klimawandel, 2nd edn. Books on Demand GmbH, Norderstedt

Garratt JR (1978) Flux profile relations above tall vegetation. Quart J Roy Meteorol Soc 104:199–211. ▶ https://doi.org/10.1002/qj.49710443915

Jäckel S, Borowiak A, Stacey B (2021) Standardization in atmospheric measurements. In: Foken T (Ed) Springer handbook of atmospheric measurements. Springer International Publishing, Cham, pp 93–106. ▶ https://doi.org/10.1007/978-3-030-52171-4_4

Kolle O, Kalthoff N, Kottmeier C, Munger JW (2021) Ground based platforms. In: Foken T (Ed) Springer handbook of atmospheric measurements. Springer Nature, Cham, pp 155–182. ▶ https://doi.org/10.1007/978-3-030-52171-4_6

Kopp G, Lean JL (2011) A new, lower value of total solar irradiance: evidence and climate significance. Geophys Res Lett 38(1):L01706. ▶ https://doi.org/10.1029/2010GL045777

Lalic B, Eitzinger J, Marta AD, Sremac AF, Orlandini S, Pacher B (2018) Agricultural meteorology and climatology. Firenze University Press, Firenze

Leclerc MY, Foken T (2014) Footprints in micrometeorology and ecology. Springer, Heidelberg, New York, Dordrecht, London. ▶ https://doi.org/10.1007/978-3-642-54545-0

Mason PJ (1988) The formation of areally-averaged roughness length. Quart J Roy Meteorol Soc 114:399–420. ▶ https://doi.org/10.1002/qj.49711448007

Mauder M, Foken T, Cuxart J (2020) Surface energy balance closure over land: a review. Boundary-Layer Meteorol 177:395–426. ▶ https://doi.org/10.1007/s10546-020-00529-6

Metzger S (2018) Surface-atmosphere exchange in a box: making the control volume a suitable representation for in-situ observations. Agric Meteorol 255:68–80. ▶ https://doi.org/10.1016/j.agrformet.2017.08.037

Muller CL, Chapman L, Johnston S, Kidd C, Illingworth S, Foody G, Overeem A, Leigh RR (2015) Crowdsourcing for climate and atmospheric sciences: current status and future potential. Int J Climatol 35:3185–3203. ▶ https://doi.org/10.1002/joc.4210

Orlanski I (1975) A rational subdivision of scales for atmospheric processes. Bull Am Meteorol Soc 56:527–530. ▶ https://doi.org/10.1175/1520-0477-56.5.527

Persson A (2017) The Story of the Hovmöller Diagram: an (almost) eyewitness account. Bull Amer Meteorol Soc 98(5):949–957. ▶ https://doi.org/10.1175/BAMS-D-15-00234.1

Raabe A (1986) Zur Höhe der internen Grenzschicht der Atmosphäre bei ablandigem Wind über See. Z Meteorol 36:308–311

Raupach MR, Finnigan JJ, Brunet Y (1996) Coherent eddies and turbulence in vegetation canopies: the mixing-layer analogy. Boundary-Layer Meteorol 78:351–382. ▶ https://doi.org/10.1007/978-94-017-0944-6_15

SI (2019) Le Système international d'unités (The international system of units), 9th edn. Bureau International des Poids et Mesures, Sèvres

Spirig C, Ammann C, Neftel A (2017) ART footprint tool. Zenodo 816236. ▶ https://doi.org/10.5281/zenodo.816236

Stull RB (1988) An Introduction to boundary layer meteorology. Kluwer, Dordrecht. ▶ https://doi.org/10.1007/978-94-009-3027-8

Trenberth KE (2022) The changing flow of energy through the climate system. Cambridge University Press, Cambridge. ▶ https://doi.org/10.1017/9781108979030

VDI (2025) Umweltmeteorologie - Meteorologische Messungen - Grundlagen (Environmental meteorology—Meteorological measurements—Basics), VDI 3786, Blatt (part) 1. Beuth-Verlag, Berlin

Wild M (2020) The global energy balance as represented in CMIP6 climate models. Clim Dyn 55(3):553–577. ▶ https://doi.org/10.1007/s00382-020-05282-7

WMO (1979) Proceedings of the world climate conference (WMO No. 537). World Meteorological Organization, Geneva

WMO (2024) Guide to instruments and methods of observation, WMO-No. 8, Volume I—Measurement of meteorological variables. World Meteorological Organization, Geneva

Methodical Recommendations for Micrometeorological Applications

Josef Eitzinger, Tamás Weidinger, Branislava Lalic, Thomas Foken, Zoltán Istenes, Britta Jänicke, and Erich Mursch-Radlgruber

B. Lalic et al. (eds.), *Micrometeorological Measurements*,
Springer Textbooks in Earth Sciences, Geography and Environment,
https://doi.org/10.1007/978-3-032-03884-5_2

This section is a methodological guide for conducting micrometeorological measurements across various land cover types. Particular attention is given to cover specific atmospheric processes, the representativeness of measurement locations, installation conditions, and sensor selection. Practical recommendations are provided to ensure reliable data collection, highlight common deviations from expected measurements, and offer strategies to address these anomalies. At the end of the section, the reader will find a short introduction to stationary, mobile, transect and crowd-sourcing measurements, and data acquisition techniques.

Straightforward navigation through this section requires familiarity with land cover types, the distinction between natural surfaces, and the Local Climate Zones (LCZ) concept, as these factors significantly influence micrometeorological measurement outcomes.

Land cover types are classified into two main categories: non-vegetated and vegetated surfaces. Non-vegetated surfaces include water surfaces (such as liquid water, snow, and ice) and bare soil (including rock). Vegetated surfaces comprise low vegetation (such as short and tall grass and shrubs) and tall vegetation (including fruit trees and forests).

Water surfaces refer to any surface covered by water, including but not limited to natural, artificial, frozen, and intermittent water bodies. Water, snow, and ice have different physical characteristics and each play distinct roles in land-surface atmosphere interaction influencing local, regional and even global meteorological conditions. For example, water moderates temperature through high heat capacity and evaporation, snow reflects to a high portion solar radiation and insolates the ground, and ice maintains high reflectivity and stability of the lower atmosphere, affecting the energy exchange processes.

Bare soil is an area of land surface that is not covered by water, vegetation, or any other type of ground cover but can be covered by rocks.

The main difference between**low and tall vegetation** is not in height but in structure, which determines aerodynamic characteristics and turbulent transfer between plant canopy and lower atmosphere. For example, maize can be taller than some fruit trees but it will be still considered as low vegetation (tall grass type) while fruit tree will be classified as tall vegetation.

Low vegetation's main characteristic is, almost, uniform leaf area density (LAI), and it is classified as short grass (up to 30 cm height) and tall grass. This type of land cover is often considered as a layer of vegetation sandwiched between soil surface and the atmosphere. It is important to have in mind that height of this vegetation can vary over the year, depending on climatic characteristics and vegetation type (especially for agricultural crops).

Tall vegetation canopy has a distinct vertical structure consisting of stem and crown area, with possible lower strata vegetation in the form of shrubs. This structure allows for specific and very complex turbulent transfer within and above plant canopy affecting micrometeorological conditions in vertical and horizontal directions.

Local Climate Zones: In an attempt to better characterize local climate, efforts have been made to describe it based on typical surface features (built-up areas, building density, vegetation, energy sources). The result of this work was a compilation of urban climate zones, which in subsequent years was expanded to include rural climate zones (Stewart and Oke 2012), so that today we speak of local climate zones (LCZ). Criteria for classification include the ratio of the hemisphere of sky visible from the ground to the unobstructed hemisphere, the mean ratio of height to width of street canyons, building spacing, tree spacing, the ratio of building floor area to total floor area, the ratio of impervious area (paved, rock) to total area, the ratio of pervious area (bare soil, vegetation, water) to total area, the geometric mean of building heights and tree/plant heights, and a classification of effective terrain roughness. For currently used numerical values of these characteristic quantities, see Oke et al. (2017). This classification system also allows for the assignment of micrometeorological measurements to specific LCZ, as detailed in the following sections: ▶ Sect. 2.4 for low vegetation (LCZ D–F), ▶ Sect. 2.5 for high vegetation (LCZ A–C), and ▶ Sect. 2.6 for urban structures (LCZ 1–10).

2.1 Some Factors Affecting Micrometeorological Conditions

Small variations in local weather conditions and land surface cover can lead to significant spatial and temporal variations of temperature, humidity, and other micrometeorological elements as well as processes (such as heat fluxes). By comprehensively analysing and understanding these factors and their interplay, one can optimize its measurement strategies and improve expected outcome.

Local microclimate depends significantly on the specific **weather conditions**. Under clear, cloudless skies, greater variations can develop on small time and spatial scale due to stronger energy exchange by radiation at the surfaces compared to overcast, rainy weather. High wind speeds and cloudiness typically result in smaller spatial differences in temperature and humidity in the surface layer. Conversely, with a clear sky, strong daytime global radiation, night-time longwave outgoing radiation, and weak to moderate wind speeds, we observe larger local differences in values of the near-surface micrometeorological elements such as air temperature and humidity, as well as radiation balance components (due to shading and different surface albedos). The mosaic of different land surface covers as well as ground/soil characteristics (such as heat capacity, soil water content) reinforces these temporal and spatial variation differences of surface temperature and soil moisture, reflected in the local surface energy budget components, evaporation, and vertical profiles of micrometeorological variables (flux-profile relationships, inhomogeneities, advection processes, Cuxart et al. 2016).

Adequate consideration of surface conditions and ongoing atmospheric processes in respect of temporal and spatial resolution is crucial for the relevance of micrometeorological measurements and application-specific. ◘ Tables 2.1 and 2.2 list six important surface condition parameters influencing surface layer processes across various land cover types (see also ▶ Chap. 1). The roughness length which reflects the aerodynamic characteristics of the underlying surface, and the footprint, which includes flow and stability conditions, are important for all cover types while zero-plane displacement height should be taken into consideration in the case of vegetated surface. ◘ Table 2.1 contains the influencing factors that must be taken into account when selecting the measurement lo-

◘ **Table 2.1** Problems influencing surface layer processes above different types of surfaces

Influencing parameter	Section	Water	Bare soil	Low vegetation	High vegetation	Complex surfaces
Footprint	1.2.4	(–)	(–)	(–)	(–)	+
Zero-plane displacement height	1.2.3	–	–	+	+	+
Obstacles	1.2.5	(–)	(–)	(–)	(–)	+
Internal boundary layers	1.2.5	(–)	(–)	(–)	(–)	+
Heat sources	2.3	–	–	–	–	(+)

Legend: – not relevant, (–) not relevant if the surface is uniform within the footprint, otherwise relevant, + relevant, (+) may be relevant—ask specialist

2

Table 2.2 Atmospheric processes influencing surface layer structure and development above different types of surfaces. Legend is the same as in Table 2.1

Atmospheric processes	Section	Water	Bare soil	Low vegetation	High vegetation	Complex or urban surfaces
Stable stratification	1.2.6	+	+	+	+	+
Convection	2.2	–	–	–	(+)	(+)
Coherent structures	2.2	–	–	–	(+)	(+)
Low-level jet	[a]	–	–	–	(+)	(+)
Afternoon transition	1.2.7	–	+	+	+	+
Advection	[b]	(–)	(–)	(–)	(–)	+
Boundary layer height	1.2.1	–	–	–	(+)	(+)

[a] During stable stratification, the wind at ground level frequently becomes lighter or calm at night, and the wind aloft may accelerate to high speeds in a phenomenon called low-level jet. This can have an influence on measurements near the ground—clarification by experts
[b] Transport of atmospheric property by mass motion (velocity field). These advected properties can influence the measurements, especially in heterogeneous terrain, e.g. by generating an oasis effect (see ► Sect. 1.2.7)—clarification by experts

cation, and Table 2.2 contains atmospheric processes that must be taken into account specifically when analysing the data and developing the measurement design.

The **spatial and temporal representativeness** of measurements is always given if the surface is as homogeneous as possible and uniform in the footprint area. The same applies if the measurement variable changes as little as possible spatially. For further details, see ► Sect. 1.2.

2.2 Water Surfaces

Measurements over water pose a particular challenge. This applies above all to the choice of measurement platform. Buoys, floats (pontoons), and masts fixed to the ground are usually used. It is also common to set up masts on sea bridges or on the shoreline (Kolle et al. 2021). The measuring devices must not be installed too close to the surface of the water due to wave action. This restricts gradient measurements. If possible, sonic anemometers should not be used on movable platforms without a sufficient stabilizing body under water (buoys, floats), unless inclinometers have been installed for angle correction. The use of cup anemometers is unproblematic, as they still provide the correct measured value for the wind speed even at tilt angles of up to around 20° (Foken and Bange 2021). Installation close to the shore requires particular care when determining the footprint in order to exclude measurements that are partially influenced by the land.

2.2.1 Behaviour of Meteorological Variables Over Water

Meteorological variables do not behave fundamentally differently over water surfaces than over land surfaces with low vegetation. However, due to the low albedo and high heat capacity, water is a large heat reservoir, which leads to higher temperatures at night compared to land surfaces. The opposite occurs during the day. The lower surface roughness leads to higher wind speeds. The differences between land and water surfaces can be seen most impressively in sensible and latent heat flux. As a rule, evaporation over vegetated land areas in the temperate climate zone is higher than over lakes and rivers, at least as long as the water temperature during the day is lower than the air temperature. Stable stratification then prevails over lakes (negative sensible heat flux) and evaporation is reduced. At night, the temperature conditions are usually reversed and, with unstable stratification, water can evaporate more over the lake than during the day; only the lower wind speeds have a reducing effect (Beyrich et al. 2006). This is illustrated in ◘ Fig. 2.1 for a lake in comparison to an agricultural field. It is important to note that water reservoirs with sufficient water depth evaporate significantly less than land areas, even in warm weather periods.

2.2.2 Influences of Waves and Water Depth

Waves significantly determine the surface roughness. It is not wave height and wave length that are decisive, but the steepness of the waves. Thus, a water surface appears smoothest with wind waves at a wind speed of around $4\,m\,s^{-1}$. At lower wind speeds, there are very steep capillary waves (rippling of the surface). The water depth also has an influence on the waves. The shallower it is, the steeper the waves become and the exchange increases and must be taken into account with additional correction factors in determining equations (Panin and Foken 2005; Foken and Mauder 2024).

2.2.3 Influences of the Footprint

Since roughness and heat capacity differ significantly between land and water surfaces and therefore also the turbulent fluxes, exact knowledge of the footprint is very important. This is especially true for measurement locations close to the shore or on the shoreline. In the latter case, the increased exchange in the shallow water area may have to be taken into special consideration. Measurements should therefore be excluded if the majority of the footprint (>90%) is not above the water surface. Partial modelling can be used to obtain continuous results for the water surface. If the footprint is clearly above the water or the land, a model can be adapted in each case, which can then be used to replace incorrect data due to insufficient footprint (Biermann et al. 2014).

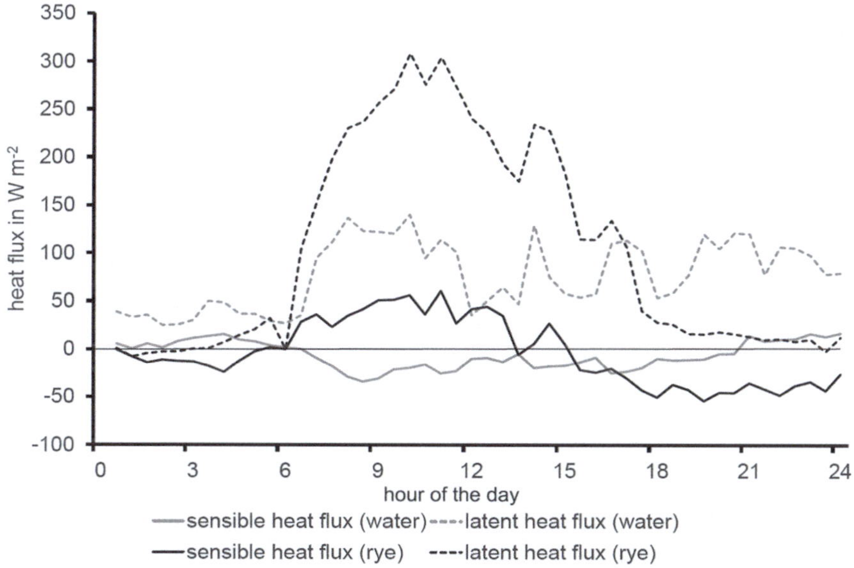

◘ **Fig. 2.1** Typical diurnal variation (0–24h) of the sensible (solid line) and latent heat flux (dashed line) over a rye field (black lines) and a water surface 3 km away (grey lines) on 25.05.2003 during the LITFASS-2003 experiment effect (Beyrich et al. 2006). The oasis effect over the rye field is clearly recognizable. From Foken and Mauder (2024), with kind permission of Springer

2

2.3 Soil and Soil Surface

Soil is the top, soft, fertile layer of the Earth's surface, which provides crucial ecosystem services and impacts the energy and water cycle or balance significantly. The soil surface is the lower boundary for atmospheric processes and the upper boundary for soil processes such as heat, water, and nutrient transport. The top 2 m of soil is the most important reservoir of water and nutrients for vegetation.

The exchange of energy, water vapour, and momentum between bare soil and the atmosphere depends on the soil surface's roughness, albedo, soil heat capacity, and conductivity. These soil characteristics are determined by soil type and structure, as well as the humidity of the soil. For example, sandy soils and deserts are characterized by a higher albedo (above 0.3) than vegetation (mostly in the range of approx. 0.18–0.23) and low thermal conductivity, particularly under low soil moisture. Precipitation and evaporation are the main water balance components determining soil moisture and energy and the water balance of the topsoil layer and the atmospheric surface layer above it.

Vertical profiles of meteorological elements vary above different soils during the day due to differences in turbulent transfer originated by specific roughness elements (share-related or mechanical turbulence) and thermal characteristics of the surface (buoyancy-related or convective turbulence). In general, low intensity of turbulent transfer leads to a more stable boundary layer and lower variation of meteorological elements.

2.3.1 Soil Characteristics Determining Energy and Water Fluxes and Budgets

Soil compartments consist of bulk soil, litter, root, and the rhizosphere, all of which provide plants through its roots directly with water and nutrients. There are many interactions between plants, soil compartments, and soil-born organisms whose diversity and amount are extensive. Within this complex system of interactions, soil stores, binds, filters, and transforms various substances important for soil fertility and ecosystem services, such as purification of water. In agricultural soils, the type of soil, including its solid particles (organic and mineral) and their size distribution (known as soil texture), determines both the total soil pore volume and the distribution of pore sizes (◘ Table 2.3).

Soil pore volume can be calculated as follows:

$$PV = \left(1 - \left(\frac{BD}{SD}\right)\right) \cdot 100 \tag{2.1}$$

where PV is soil pore volume (% vol.), BD is bulk density (g cm^{-3}), and SD is specific mass of the solid soil particles (g cm^{-3}).

The soil hydraulic characteristics influence all basic elements of the terrestrial hydrological cycle at all spatial and temporal scales. **Soil water balance and dynamics** (see ► Sect. 1.1.5) is highly sensitive to soil texture, soil structure, and soil pore volume and size distribution. Dependent parameters are **soil water holding capacity, infiltration, sur-**

◘ **Table 2.3** Portion of solid soil particles from total soil volume (inverse to total pore volume)

Soil type	Portion of solid soil particles (%)
Less organic, mineral soils (typical for arable soils)	approx. 50
Soils with higher organic content (typical beneath forests, grassland)	40–45
Soils with very high organic content (i.e. swamps)	10–35
Compacted mineral soils	60–65
Sandy soil	44–64
Silt soil	45–70
Clay soil	30–65

face runoff, evapotranspiration, drainage, capillary rise, and soil water conductivity (Ács et al. 2010).

The water flow into the soil starts with the infiltration rate from the soil surface, which depends on the water flow and the related hydraulic conductivity within the soil medium itself. In a simple approach (one-dimensional), the water flow amount and speed can be calculated according to Darcy's law:

$$Q = k\frac{d\Phi}{dl} \tag{2.2}$$

or

$$v = k\frac{h}{l} \tag{2.3}$$

where Q is the water flux amount (in cm^3 cm^{-2} s^{-1}), k is an empirical flux constant, v is the speed of water flux (in cm s^{-1}), dΦ is the water pressure difference, h is the hydraulic pressure, and dl is distance between two points in the soil.

According to Darcy´s law, the large range of water flow and related infiltration capacities of soils impacts characteristics that are relevant for the whole water balance (▣ Table 2.4).

A key characteristic for estimating soil water balance, total soil water-holding capacity, and the plant-available water is the relationship between soil water content and soil water potential (expressed as a suction, negative pressure), also known as **pF-curve** (▣ Fig. 2.2). This relationship is strongly influenced by the soil's pore size distribution and related capillary forces.

In Agrometeorology, three critical indicators of crop water availability, which depend on the soil type, are the following: the soil water content at the permanent wilting point, which is the approximated lower limit of water uptake by plants; the field capacity—the water amount held in the soil against gravity, which indicates good water supply for crops; and the saturation water capacity, where a soil is fully water saturated, which occurs, e.g. after heavy rain.

For the total amount of plant-available water and the related transpiration potential (which is the major portion of latent heat flux above active vegetation), an additional dynamic, biotic factor is of crucial importance—the **effective rooting zone** or depth. This determines the volume of soil (and the related amount of available water) accessible and useable by the plant roots. The rooting zone varies based on plant type, phenological status, genetically determined rooting systems, and on the soil conditions, which may act as forcing or limiting factors for root growth, making it highly dynamic in both time and space. Since this information is often absent from soil maps, soil-crop-water balance estimates (e.g. by crop, ecosystem, land surface models) remain highly uncertain, if no credible in situ data on these variables are available. This is particularly relevant for crop field-based assessments in agriculture, where soil management and cultivation may further modify root growing conditions. Remote sensing methods can also provide useful data related to spatial variations

▣ **Table 2.4** Range of soil water flow characteristics of soils (Lalic et al. 2018)

Flux constant, k	Water conductivity in soil (in cm per day)	Flux time needed for amount of 10 mm (= 100 m^3/ha)	Water conductivity (porosity)
10^{-1}	8640	10 s	High—fast flux
10^{-2}	864	1 min 40 s	
10^{-3}	86	16 min 40 s	
10^{-4}	8.6	2 h 46 min	Medium
10^{-5}	0.86		
10^{-6}	0.086	11 days	Low—slow flux
10^{-7}	0.0086		
10^{-8}	0.00086	1 year	

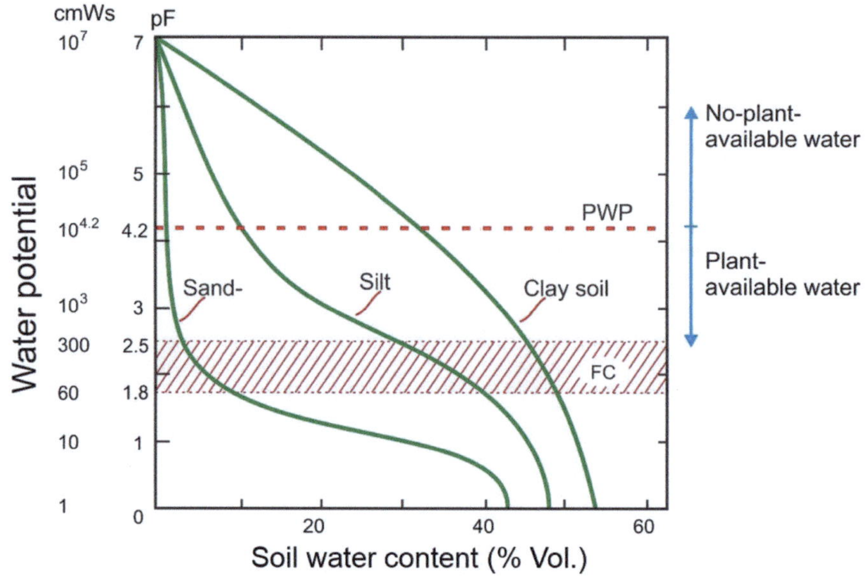

Fig. 2.2 Scheme of the relationships of volumetric soil water content (% Vol.) and water potential of three main soil types (pF-curves). The pF value is the logarithmic expression of cm WS (water column in cm), PWP is the permanent witling point, and FC is the field capacity. After Lalic et al. (2018), © Authors, CC BY 4.0

of, e.g. actual transpiration, canopy temperatures, drought stress conditions, spatial yield variations, and the like.

Beyond the water flux from the soil and/or vegetation surface into the atmosphere (latent heat flux), another important component of the **surface energy balance** (see Note 1.7 in Sect. 1.1.4) is the **soil heat flux** (see Fig. 1.11 in Sect. 1.1.4). This flux is smaller than the sensitive and latent heat fluxes together, and in an overall range of **maximum about 25%** of the net radiation, depending on the soil and soil surface conditions as well as the season and time. For example, during the spring season when soils are warming up, it is about 10% of the net radiation (energy transported into the soil) and vice versa in the autumn. Soil heat flux is changing its direction also over the day-night cycle, influenced by the overall climate and weather conditions. **Soil heat flux generally balances out over the course of a year**. Further the soil surface conditions, especially isolating soil cover such as organic mulch, snow layer, and crop canopy, can strongly reduce soil heat flux. Within the soil, the heat flux is determined by the thermal conductivity and heat capacity of the soil materials, determining thermal diffusivity of different soil layers. These soil properties are also influenced by dynamically changing soil moisture.

The soil water balance factors (see Note 1.11 in Sect. 1.1.5) are important for many model applications involving soil-plant-atmosphere interactions. The above-discussed soil physical properties, for example, influence the heat flux into or out of the soil and the evapotranspiration (latent heat flux) of soil and vegetation. On a larger spatial scale, the ratio and values of sensible and latent heat fluxes shape the convective processes, i.e. the height of the boundary layer and, given sufficient energy and moisture, the formation, location, and amount of convective precipitation (Breuer et al. 2012; Ács et al. 2014; Göndöcs et al. 2015).

Due to the high spatial (in fact, three-dimensional) variation of soil properties, many uncertainties arise when using data from soil maps, which are typically based on single-point profile estimates, subjective descriptions, and interpolations. These uncertainties have to be kept in mind for any model application, whether it's for crop yield predictions, precision farming applications, or larger scale spatial applications such as land surface models, which are also part of weather forecasting models. In recent years, remote sensing data—from drones and satellites—has been used to complement traditional soil data, reducing some uncertainties by monitoring surface characteristics like soil wetness, temperature, and spectral reflectance. However, these techniques are still limited, as they only capture surface conditions through spectral emissions or reflections, providing no information about subsurface properties.

2.3.2 Behaviour of Meteorological Variables Above Bare Soil

The **air temperature above bare soil** during the day is highest near the soil surface due to the solar radiation absorption by the soil, leading to ground heating. Air temperature decreases with height. The rate of decrease, or lapse rate, depends on surface characteristics. For example, above dry soils with low heat capacity it is higher due to intensive surface heating. Above wet soils with higher heat capacities the lapse rate is lower.

Additionally, daily air temperature variation is higher than when vegetation is present. The air temperature is lowest near the surface at night due to intensive radiative cooling. In this case, temperature increases with height, leading to temperature inversion and a stable atmospheric layer formation. This inversion traps cooler air near the surface, with all sunset time air humidity, allowing dew formation when the surface temperature reach the dew point.

The **soil surface temperature** depends on the heating of the surface layer and the soil and site characteristics (such as soil wetness or topography) affecting energy transfer. Heating and soil characteristics both change during the day and year, depending on weather and climate conditions, and produce the diurnal and annual cycles of the soil temperature profile.

During the day, the surface soil layer attains its maximum temperature approximately one hour after maximum solar radiation, while its minimum is reached just before sunrise. However, a time lag in maximum and minimum soil temperatures increases with soil depth. That time is required for surface layer heating and energy transfer through the soil column, which greatly depends on the heat capacity of the soil. Daily variation in soil temperature decreases with depth until it reaches a constant temperature level at some depth (◘ Fig. 2.3). The depth of this level depends on soil type and wetness, season, and latitude. The daily course of soil temperature has a seasonal pattern, particularly evident at the surface layer (◘ Fig. 2.4). Soil temperature variation with depth during the night mimics air temperature profile—soil temperature increases from the soil surface towards constant-temperature depth.

Air humidity of a thin layer just above bare soil strongly depends on soil moisture. If there is no moisture deficit, a common approximation is that air just above the soil surface is in saturation on temperature, which equals soil surface temperature. During the day, air-specific humidity variation with height is a result of an interplay between the intensity of soil evaporation and the efficiency of turbulent transfer. Suppose the humidity supply during the daytime is constant, and relative humidity is determined only by turbulent transfer and vertical temperature changes. In that case, we can expect that it increases with height due to a decrease in air temperature. During the night, in contrast, the relative air humidity would be highest near the ground, decreasing with height.

Wind speed above bare soil increases with height according to the logarithmic profile due to the reduction of friction between the air and the ground with height. Wind speed is typically low near the surface during nighttime and slowly changes with height.

2.3.3 Profiles and Fluxes—Bare Soil

Vertical mixing of the atmospheric boundary layer starts with processes in the roughness sublayer, within a few centimetres above the surface of bare soil. These processes are strongly affected by surface characteristics: (a) roughness scales affecting share-related turbulence production and (b) albedo, heat capacity, and conductivity affecting buoyance-related turbulence.

2

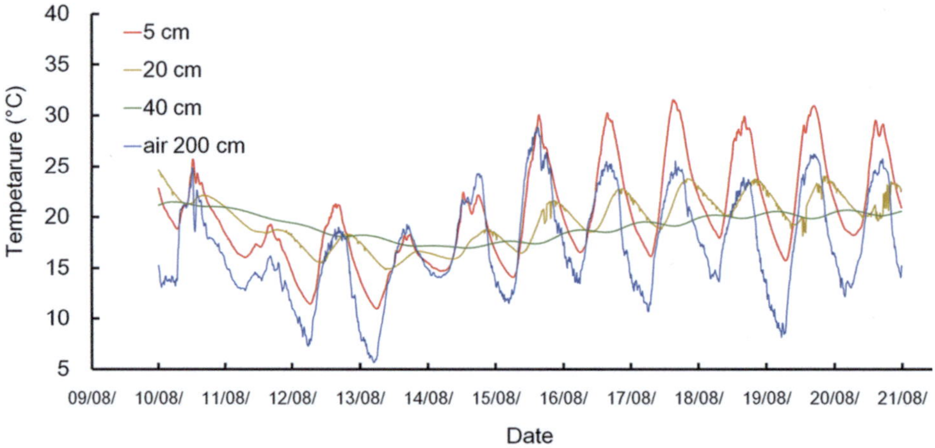

Fig. 2.3 Daily variation in soil temperature (grass cover) and air temperature during the summer 2016 in Goggendorf (Austria). *Source* BOKU-Met, after Lalic et al. (2018), © Authors, CC BY 4.0

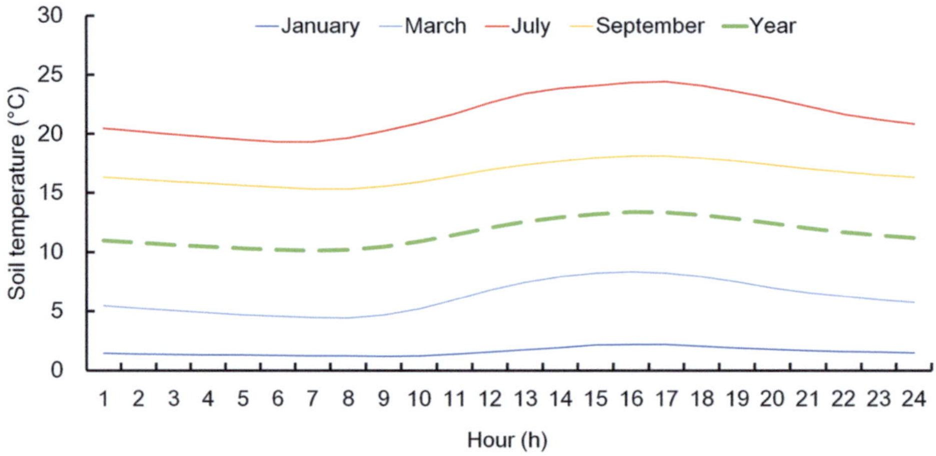

Fig. 2.4 Seasonal variation in the daily course of soil temperature at 10 cm depth in Ridjica (Serbia) (2013–2017). *Source* PIS Vojvodina, Serbia, © Authors CC BY 4.0

Bare soil surfaces with minimum roughness elements will contribute less to share-induced turbulence. Examples of low roughness surfaces are levelled and smoothed farmland typically after sowing and harvest of field crops; smooth, sandy, or silty desert surfaces; and dry, flat lake beds in arid regions. On the contrary, high roughness surfaces characterized by features that create significant friction and contribute to intensive turbulence formation are rocks, ploughed or cultivation dam (e.g. potatoes) cropland, terrain with significant erosion, and construction zones on bare soil. Dry soil surfaces with low albedo, low heat capacity, and conductivity heat up quickly, promoting convective turbulent transfer.

From the sunrise, solar heating of the soil surface creates convective turbulent transfer characterized by strong vertical mixing. The presence of wind will add up to turbulent eddies formation, enhancing momentum, energy, and water vapour exchange between the soil surface and the atmosphere. Under clear-sky weather, maximum turbulent mixing typically appears at approx. 3 pm. After that time, the intensity of turbulence slowly reduces towards the night-time regime. After sunset, ground cooling due to negative thermal radiation balance leads to stable atmospheric conditions with suppressed vertical mixing.

From the measurements point of view, it is important to keep in mind that the difference between daytime and night-time turbulence transfer leads to significant changes in the daily variation of all micrometeorological elements: For example, two maxima (approx. 9 am and 9 pm) and two minima (approx. 4 am and 3 pm) in the daily regime of water vapour pressure in continental mid-latitudes during the summer. What causes this phenomenon? In the morning, just before sunrise (approx. 4 am), the energy balance of the Earth's surface and the air temperature are at their daily minimums, leading to a minimum of evaporation and water vapour supply to the atmosphere. As incoming energy increases, the intensity of evaporation and the amount of water vapour in the air above the evaporating surface also increase. During the first few hours after sunrise, when the vertical transfer is still relatively weak, water vapour accumulation occurs, increasing its pressure to the maximum, usually at about 9 am. At midday, the surface and air warms up significantly, thereby enhancing the vertical (turbulent) transfer of water vapour and reducing it to the minimum level, coinciding with the maximum atmospheric turbulent transfer in the early afternoon (at about 3 pm). As vertical transfer decreases towards sunset, accumulation of water vapour retakes place, producing the second maximum of water vapour pressure (at about 9 pm), even if the intensity of evaporation drops due to a decrease in incoming radiation and surface and air temperatures. More details can be found in Lalic et al. (2018).

2.4 Low Vegetation

Vegetation presence on the land surface affects its roughness, heat capacity, and conductivity. It also introduces new sources and sinks of energy, water vapour, and momentum into land surface-atmosphere exchange processes. Low vegetation is typically considered as a layer of porous material on the ground surface, which serves as a significant source and sink of atmospheric humidity, heat, and momentum. When preparing measurements in low vegetation, it is important to account for differences in the vertical profiles of specific humidity, temperature, and wind speed within and above short vegetation.

Biosphere-atmosphere interaction processes above short grass differ significantly from those above tall grass. Short grass behaves as a vegetation layer on the ground surface with negligible depth. From the perspective of atmospheric processes, short grass is very similar to bare soil but has a much more complex energy and water balance partitioning, including evapotranspiration and water interception. The roughness of a surface covered uniformly with short grass is lower than that of a typical bare soil surface. The heat capacity of a plant canopy depends on canopy density and the climate seasons (as leaf area density can change significantly throughout the year depending on climate zones). Depending on the underlying soil, the canopy can either increase or decrease heat capacity of the surface.

Compared with chernozem soil, the heat capacity of a short grass canopy is generally lower. Chernozem soil, known for its high organic matter content and moisture carrying capacity, typically has a higher heat capacity because water has a high specific heat capacity. In contrast, a short grass canopy, consisting primarily of plant material and air space, has a lower heat capacity because plant material and air have lower specific heat capacities compared to water. Therefore, a short grass canopy heats up and cools down more quickly than chernozem soil. In the case of sandy soil, a similar reasoning leads to the conclusion that the heat capacity of a short grass canopy is generally higher than that of sandy soil.

Considering the aerodynamic and thermal characteristics of short grass, it is evident that its presence changes micrometeorological profiles above bare soil. Therefore, the following section is devoted to the momentum, water vapour, and energy exchange between tall grass and atmosphere and its impact on meteorological elements above and within canopy.

2.4.1 Behaviour of Meteorological Variables

During the course of the growing season leaf area above the ground changes, which can be expressed by indicators such as leaf area density *(LAD)*, leaf area index (LAI) or plant area index (*PAI*), see below, and height (*h*) of the "tall" grass canopy type. Consequently, its roughness length (approximately 0.08 h), zero plane displacement height (0.6–0.8 h), as well as drag coefficient and heat capacity, are changing as well during the growing period. Although the vertical structure of the canopy can typically be neglected, vertical profiles of all meteorological elements differ significantly above and within the canopy during the course of the day and night cycle (Fig. 2.5) and further modified by the weather conditions (especially wind).

- **LAI** (Leaf Area Index) represents the total area of leaves of a plant canopy above 1 m^2 of ground taking only one side of each leaf into account. It is expressed as a dimensionless quantity (leaf area per unit of ground area, $m^2\ m^{-2}$)
- **LAD** (Leaf Area Density) represents the vertical and horizontal canopy structure and is defined as the total one-sided leaf area per unit volume ($m^2\ m^{-3}$).
- **PAI** (Plant Area Index) represents the total area of plant surfaces (e.g. including stems and leaves, needles) above 1 m^2 of ground. It is expressed as a dimensionless quantity (plant area per unit of ground area, $m^2\ m^{-2}$).

If the variation of the vertical structure of canopy is more pronounced (such as in forests), LAD(z) represents leaf area within a canopy volume of 1 m^2 of base and the vertical distance dz between heights z and $z+$dz. If the vertical distribution of leaves is known over the canopy profile, LAI can be calculated as

$$\mathrm{LAI} = \int_0^h \mathrm{LAD}(z)\mathrm{d}z, \tag{2.4}$$

where h is canopy height. In case of canopies of relatively uniform structure, i.e. constant LAD(z), LAI = LAD $\cdot\ h$.

Radiation intensity, during the day, decreases intensively within the canopy due to solar radiation absorption by leaves and stems, reaching the ground in closed canopies with

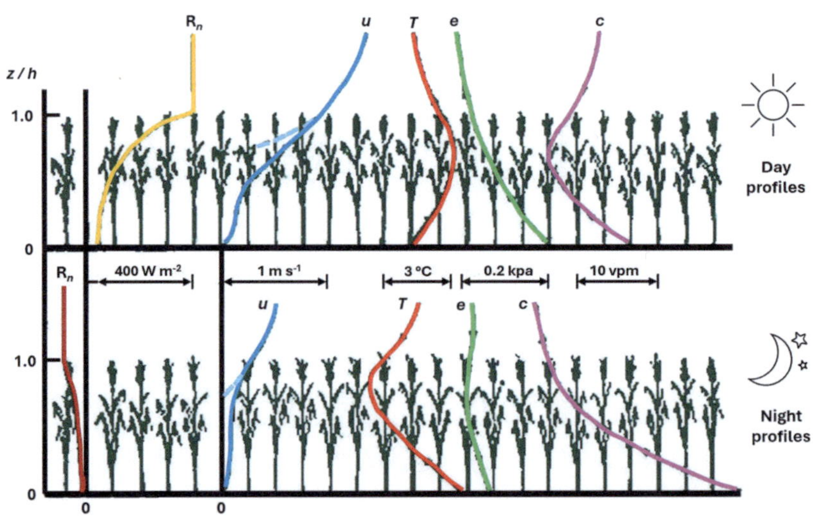

Fig. 2.5 Idealized profiles during cloudless day (top) and night (down) of net radiation ($R_n = -Q_s^*$), wind speed (u) with extrapolated profile, air temperature (T), vapour pressure (e), and CO_2 concentration (c) in a homogeneous crop field (maize) of height h plotted as a function $\frac{z}{h}$, where z is the height above the surface and h is the height of the canopy. From Monteith and Unsworth (2014), with kind permission of Elsevier (modified)

only about 10% or less of its value at the top of the canopy. The major portion of absorbed energy causes increase of plant and canopy air temperature, as well as evapotranspiration (latent heat flux) and sensible heat flux from the canopy to the atmosphere. This reduces the air temperature lapse rate above vegetation compared to bare soil. During the night the highest negative net (thermal) radiation can be found at the top of the canopy.

Air temperature above vegetation decreases with height during the day, but within the vegetation, it increases from the ground to the canopy top due to intensive radiation absorption. This forms a typical air temperature profile with the maximum temperature occurring approximately at the maximum LAD height. In an uniform canopy, the highest temperature is found in the top 30% of the canopy. During night, the highest negative thermal balance at the top of the canopy leads to the lowest leaf and air temperatures at this level.

Air humidity, during the day, decreases with height above the canopy as the distance from the plant canopy, which is the source of water vapor, increases. The rate of decrease depends on the intensity of turbulent transfer, available water, plant variety, and growth stage. Within the canopy, humidity is highest near the ground and decreases with height. During the night, humidity variation is negligible due to very low evapotranspiration intensity and a typically stable atmosphere.

Wind speed above vegetation increases with height (following a logarithmic low) because distance from the rough vegetation top increases. The wind speed lapse rate above vegetation is typically lower than above bare soil, but it is strongly affected by canopy structure and density. A dense canopy with fully developed leaves has smaller roughness and produces a lower lapse rate, particularly in the roughness layer immediately above short vegetation. Within the vegetation canopy, there is an intensive reduction of wind speed (following a hyperbolic cosine low) up to the lower half of the canopy, where almost all momentum is dissipated and only a small fraction reaches the ground. During the night, the wind profile above and within vegetation follows daily patterns but with typically lower wind speed values.

CO_2concentration profiles vary greatly during the day and over the season due to changes in the intensity of the physiological processes, such as photosynthesis and respiration, which contributes to the sources and sinks of CO_2. Soils permanently emit CO_2 depending on soil organisms and plant root respiration activity where it can accumulate above the ground and within canopies during calm conditions (which are in general more expressed during night). During photosynthetic active periods, however, plants can reduce CO_2 concentration in the canopy below the free atmosphere level due to stomatal CO_2 uptake (depending on the wind velocity).

2.4.2 Profiles and Fluxes—Low Vegetation

Turbulent transfer above and within low vegetation exhibits specific characteristics influenced by the vegetation structure, height, and density. Above the canopy, turbulent transfer is typically enhanced due to interaction between wind and the roughness elements such as leaves and stems. This creates turbulent eddies that mix air more effectively. Eddy diffusivity is lower within the canopy compared to above it. The dense structure of the vegetation dampens large eddies and restricts mixing. It affects energy and water balance of the canopy as well as overall microclimate of the atmosphere within and above canopy.

2.5 High Vegetation

The following section is mainly based on Foken and Mauder (2024): The description of meteorological processes over high vegetation is different from that for dense vegetation (e.g. tall grass), which was as a porous layer with a vertical displacement of the height (zero-plane displacement) in the profile equation. In a forest, the canopy and the understorey are regions where complex micrometeorological processes occur. Therefore, the crown is a layer that often decouples the understorey from the atmosphere above the forest canopy. The energy and matter exchange between the atmosphere and the upper crown is often completely different from those between the lower crown and the trunk space and the soil. The stratification in both layers can be different, and the gradients in each layer can be of opposite sign. This has a significant influence on measurement and modelling of exchange processes, which are very complex (Finnigan 2000; Monson and Baldocchi 2014).

2.5.1 Behaviour of Meteorological Variables in a Forest

The typical daily cycle of the temperature in a forest is illustrated in Fig. 2.6. The maximum air temperature occurs in the upper crown usually about 1–2 h after local noon. Below the crown, the daytime temperatures are lower. During the night, the minimum temperature occurs in the upper crown due to radiation cooling. Because the cool air in the crown drops to the ground surface, the minimum of the temperature near the ground occurs a short time later. Especially in the evening, it is warmer in the forest than in the surroundings. Fields near the forest edges may be damaged by frost in the morning due to the outflow of cold air, especially on slopes.

Thus, during the day it remains pleasantly cool in the trunk room and on forest paths, usually a few degrees colder compared to the forest edge. Although at night the temperature in the crown and in the trunk space is almost identical, the forest is perceived as pleasantly warm compared to the forest edge. The reason is the longwave radiation balance in the forest. Since the forest floor and the crown space have almost the same temperature, no additional cooling takes place and the longwave radiation balance compared to the body surface temperature is also low, so that one does not feel any cooling effect. Calm wind supports this sensation.

The horizontal distribution of meteorological elements in a forest compared to a clearing was studied by Hübner et al. (2014) and is shown in Fig. 2.7. The shortwave radiation components K↓ and K↑ (Fig. 2.7a, b) are very low in the forest, but sunny spots can be seen at different locations and times. In the morning the shadow of the forest edge can be seen beginning at 05:30 CET on the complete clearing, decreasing from run to run. At 08:30 CET, the sunlight on the clearing is no longer influenced by the forest edge. During the day the longwave downwelling radiation L↓ (Fig. 2.7c) in the forest is almost the same as the longwave upwelling radiation L↑ (Fig. 2.7d) at the clearing. So, the surface temperature at the clearing is almost the same as the temperature of the top of forest canopy. In the night, the clearing shows significantly lower values for L↑ than in the forest. Only where the forest canopy has open areas (see the sunny spots in shortwave radiation measurements) is L↑ in the forest almost the same as L↑ at the clearing. In Fig. 2.7e, the complete

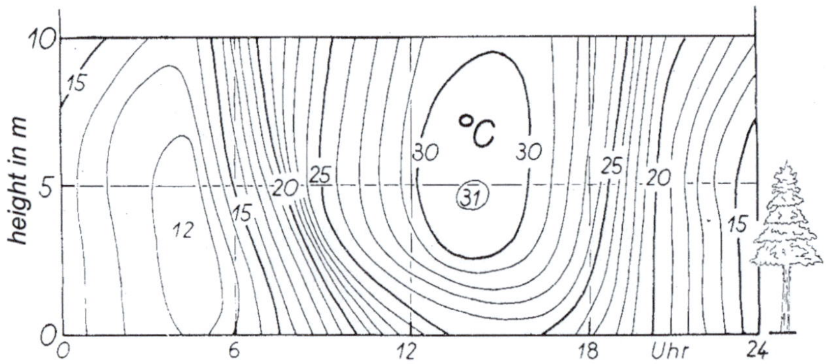

Fig. 2.6 Daily cycle of the air temperature in and above a forest according to Baumgartner (1956); published with kind permission of © German Meteorological Service (DWD), Offenbach

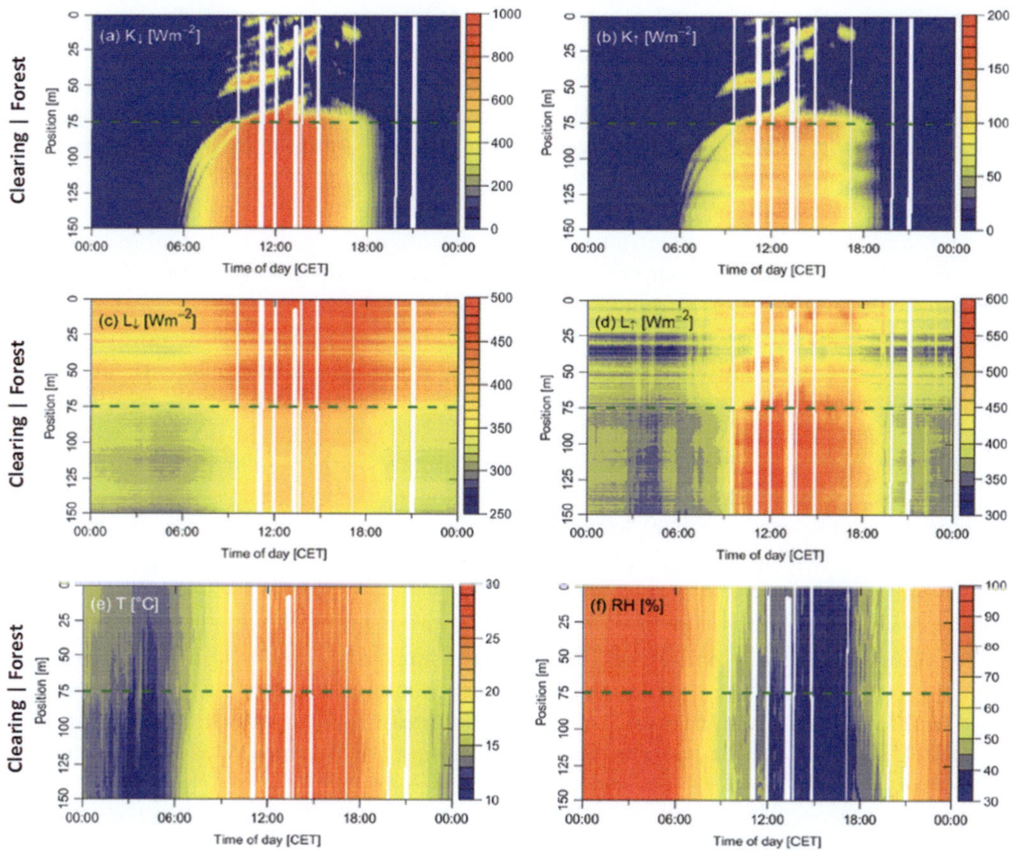

Fig. 2.7 Measured horizontal profiles at FLUXNET Waldstein-Weidenbrunnen site for 28 June 2011 at 1 m height in the trunk space (0–75 m) and a clearing (75–150 m) of shortwave down- and upwelling radiation $K\downarrow$, $K\uparrow$ (**a**, **b**), longwave down- and upwelling radiation $L\downarrow$, $L\uparrow$ (**c**, **d**), air temperature T (**e**), and relative air humidity RH (**f**). Remark: the colour scaling is different in all graphs. Taken from Hübner et al. (2014) with kind permission of © Authors, Creative Commons Attribution 3.0 License

horizontal profile for the air temperature is shown. The lowest temperatures occur during night-time and the highest during daytime at the clearing. The increase in air temperature in the morning starts at the clearing significantly earlier than in the forest, and also the decrease in the evening starts earlier at the clearing. The relative air humidity (Fig. 2.7f) is, at night-time, higher on the clearing than in the forest and at daytime this is reversed. The decrease after the sunrise is significantly earlier at the clearing than in the forest. Also, the increase in the evening starts earlier at the clearing.

The temperature structure in a forest has an absolutely different stratification than commonly observed above low vegetation. During daytime above low vegetation and above the crown, unstable stratification is typical, and during the night stable stratification is typical. Due to the relatively cool forest ground during daytime and warm forest ground during the night-time, inside the forest the opposite stratifications are realized.

A typical phenomenon of the exchange processes in and above a forest is a decoupling due to the crown. Not only the crown space but also the wind shearing above the canopy can decouple the forest from the atmosphere. Therefore, the exchange processes between the forest and the atmospheric boundary layer are occasionally intermitted. Tall plant canopies are coupled with the atmosphere by turbulent eddies most often during the daytime. Then, the energy and matter exchanges take place among the ground, the trunk space, and the atmosphere above the canopy. Reasons for the occasional decoupling between the canopy and the atmosphere are the high roughness at the upper part of the canopy, which damps the wind field, and the stable stratification in the canopy. For example, during a calm night the upper canopy is relatively warm, but cool air is stored in the understorey. A very small increase in the wind speed (gust, sweep) can cause a sudden inflow of warm air into the understorey. Also, the opposite effect (burst, ejection) often occurs during the day, when warm or moist air is suddenly ejected from the understorey. Therefore, the exchange of energy and matter within and above a forest is not a continuous process but is limited to a sequence of single events. For details, see Thomas and Foken (2007) and Peltola et al. (2021).

The wind velocity in and above a forest is greatly reduced due to friction. But in a clear trunk space with nearly no understorey growth, a secondary maximum of the wind appears (Shaw 1977; Yi 2008). This can change the momentum flux below the crown. The wind profile near the top of the canopy has an inflection point, and at forest edges internal boundary layers can develop.

2.5.2 Counter Gradient Fluxes—Coherent Structures

It is typical in the convective boundary layer that the flux—generated by convection—is not correlated to the mean, local gradient. The flux remains positive as the gradients approach zero or even slightly positive. This phenomenon is called counter gradient flux. Such a behaviour exists also in a forest. Air parcels from the trunk space penetrate through the crown into the atmosphere or a gust can penetrate into the forest. This was found to be the case not only for the fluxes of sensible and latent heat but also for the fluxes of matter, e.g. carbon dioxide (Denmead and Bradley 1985) and is schematically shown in ◘ Fig. 2.8.

While the profiles are determined by the mean structures, the turbulence fluxes contain short-time transports by gusts and bursts, which cause little change to the profiles. Relatively large turbulence eddies called coherent structures, which are not comparable with the turbulent elements of the gradient transport, make remarkable flux contributions (Holmes et al. 1996; Venditti et al. 2013).

- Coherent structures, in contrast to stochastically distributed turbulent eddies, are well organized, relatively stable long-living eddy structures, which occur mostly with regularity in either time or space.

Therefore, fluxes cannot be determined from gradients of state parameters (▶ Sect. 1.2.6). The application of either the Monin-Obukhov similarity theory or the K-approach for the exchange process at a forest site gives only meaningful results in the case of a good coupling of the atmosphere with the forest (see ▶ Sect. 2.5.1).

Above forests, coherent structures contribute about 20% to the total flux with smaller partitions at daytime and larger in the case of stable stratification at night. At the canopy, height sweeps dominate in comparison to ejections. In the trunk space and above the forest, the relationship is vice versa.

> **Note 2.1** Because of the complicated turbulence structures in and above a forest, it is strongly recommended to measure not only above the forest, but also in the trunk space.

2.5.3 Roughness Sublayer—Mixing Layer

Above a forest canopy, the profiles of state parameters are modified due to the high roughness compared with the ideal profile equations given in ▶ Chap. 1. The gradients of the wind velocity and scalars are reduced above the canopy, while the fluxes are

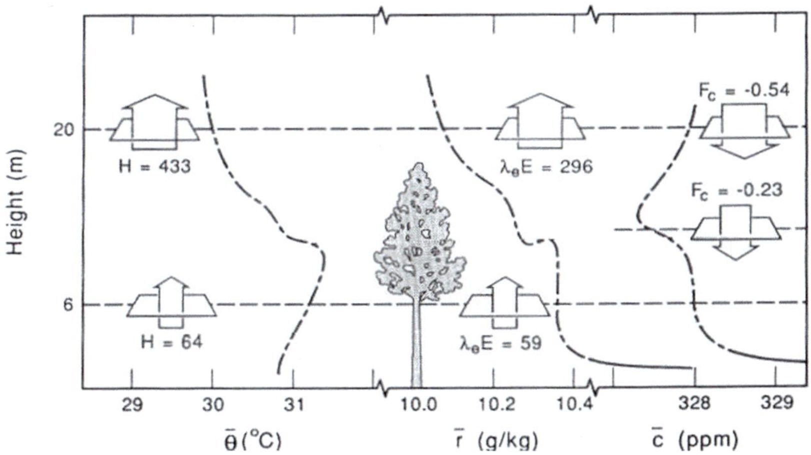

Fig. 2.8 Profile of the temperature θ, the mixing ratio r, and the trace gas (CO_2) concentration c, as well as the sensible heat flux $H=Q_H$ (in W m^{-2}), the latent heat flux $\lambda E=Q_E$ (in W m^{-2}), and the trace gas flux F_c (in mg m^{-2} s^{-1}) in a forest with counter-gradients (approximated). From Denmead and Bradley (1985), with kind permission of © Springer

increasing. This layer together with the canopy is called the roughness sublayer, which has a thickness of about three times of the canopy height. This means that above a forest and above the roughness sublayer within a distance of two times of the canopy height undisturbed conditions can be found (Thom et al. 1975; Garratt 1978). Under neutral or unstable stratification for moderately high canopies, part of the surface layer, for which the constant flux layer approach can be applied, persists above the roughness sublayer. While above low vegetation, the ratio z/z_0 for typical measurement heights has values of 100–1000, above a forest roughness sublayer this ratio is typically 5–10 (Garratt 1980).

2.6 Complex Surface Structures (for Example: Urban and Mountainous Conditions)

When conducting meteorological observations in complex structures like cities or mountainous terrain, it is crucial to considering the scale and representativeness of the measurement's location. Local Climate Zones (see introduction to Chap. 2) can help to characterize the complex and diverse urban and natural surroundings. Urban observations range from meso- to microscale and aim to quantify the urban heat island intensity at the city scale, analyse the effects of parks and new development areas at the local to microscale, and investigate detailed energy exchanges between the urban atmosphere and building envelopes or urban vegetation at the microscale.

To gather useful information in urban areas, a basic knowledge of the structure of the urban boundary layer is necessary. The urban atmosphere or urban boundary layer is typically structured in the urban canopy layer, roughness sublayer, inertial sublayer, and mixed layer (see Fig. 1.14 in Sect. 1.2.1). Most urban studies focus on the urban canopy layer (UCL). Errors might occur if the focus of a study is the urban canopy layer, which typically extends to the average building height, but individual stations are placed on rooftops of higher buildings that might enter the roughness sublayer.

2.6.1 Behaviour of Meteorological Variables in Complex Structures

Looking at the city level, typical differences in the behaviour of meteorological in cities versus rural surroundings can be observed as summarized in Table 2.5. The effect of the urban heat island with higher temperature in built-up areas is usually most pronounced during the night and often diminishes or reverses during the day.

Processes and phenomena exhibit clear distinctions between roof-level and canopy environments. Near the roof, intense wind shear and mixing prevail, while deeper within the canyon-like streets, conditions vary significantly. The canopy often offers shelter from direct wind impacts. Moreover, radiation exchange is disrupted due to the diminishing view of the sun and sky. The climate within the UCL is primarily shaped by the diverse surface properties within a small radius of a few hundred meters, with their influence diminishing rapidly as distance from the point increases (Oke et al. 2017).

Table 2.5 Typical effects of the urban climate as compared to rural surroundings (modified after Schönwiese 2024)

Radiation	
Irradiation	Lower
Sunshine duration	Lower
Air temperature	
Yearly average	Higher
During sunny days	Considerably higher
Minimum temperature	Considerably higher
Air humidity	
Relative	Lower
Absolute	Same
Fog	Lower
Clouds	
Cover	Higher
Precipitation	
Yearly average	Higher
Snow	Lower
Wind	
Average wind speed	Lower
Wind calm	Higher
Wind gusts	Higher

2.6.2 Urban Structures

Complex environments consist of various units that influence the microclimate at different scales. Depending on the specific question, different scales need to be addressed, and the arrangement of measuring devices must be adjusted accordingly.

In the **city scale**, the city itself consists of different structures and materials that contrast with those found in its rural surroundings. This contrast gives rise to distinct climatic phenomena such as the "urban heat island". Cities typically have blurred edges, making it challenging to delineate, where the urban area ends and the rural area begins. Furthermore, the impacts of urban climate extend beyond the physical borders of a city into adjacent areas, as atmospheric properties such as temperature, humidity, and air pollutants are transported downstream by the wind. Typically, phenomena observed at the city scale include urban heat islands, patterns of wind, humidity and precipitation affected by the city (Oke et al. 2017).

Neighbourhoods and block units encompasses various land use zones typically utilized in urban planning, including industrial, residential, commercial, major parkland, and undeveloped land (Oke et al. 2017). However, describing areas solely based on land use does not offer significant insight into climatic considerations, as outlined in the Local Climate Zone (LCZ) scheme. The urban block unit is defined by the layout of the road network, typically comprising adjacent street canyons with similar structures (Oke et al. 2017). In downtown areas, a block may consist of a single large building. Different road patterns may lead to blocks of various shapes, such as triangles or crescents. Block-level phenomena are influenced by specific building clusters such as shopping centres, apartment complexes, institutional buildings, and factories. These units create a mosaic of climates within the city. They are large enough to generate localized airflow patterns (e.g. in parks) or influence cloud formation (e.g. around factory complexes) (Oke et al. 2017). At this scale, effects such as local breezes or local and microscale climate variability can be studied.

The **urban canyon, or street canyon** unit, refers to the structure formed by the typical arrangement of a street and the buildings flanking it (Oke et al. 2017). While it is most prominently observed in densely built central areas of cities, its fundamental form is often replicated throughout the urban landscape. The defining characteristic of an urban canyon is its geometric shape. Canyons are typically described by their two-dimensional cross-section, denoted by the dimensionless ratio $H = W$, where H represents the height of the buildings adjacent to the street and W denotes its width. This ratio, known as the canyon aspect ratio (λ_s), holds significance as it influences various aspects of urban climates, including radiation exposure, shading effects, wind patterns, thermal comfort, and the dispersal of vehicle pollutants (Oke et al. 2017). At the level of street canyons, human bioclimate, street vortex, shadowing, and other aspects can be observed.

Facets and roughness elements encompass buildings, trees, and building lots, each composed of smaller units known as facets (e.g. roofs, walls, lawns, and paths), often distinguished by their directional orientation (such as compass direction, vertical, sloped, and horizontal; Oke et al. 2017). The geometric arrangement of an element and its roughness, radiative properties, thermal characteristics, and moisture content are all significant because each contributes to the creation of its own microclimate and influences the surrounding environment. Trees also play a crucial role in creating microclimates (Oke et al. 2017). This influence is manifested through the generation of turbulent vortices and wakes, thermals or pollutant plumes, and spatial variations in temperature and rainfall on the ground due to shading and wind effects. The detailed examination of the climatic impacts of building architecture falls within the domain of building climatology (Oke et al. 2017). Typical research questions relate to shadowing, flux exchanges, and relations between indoor and outdoor climate or dew and frost patterns.

2.6.3 Topoclimate

In complex terrain and heterogeneous landscapes, it is important to consider the landscape features for finding an appropriate measurement site. Basic knowledge of micro- and topoclimatology is required to consider influencing factors as described, e.g. by Barry and Blanken (2016). The most prominent effects are temperature gradients with elevations, expositions, and slopes, but also land cover, soil properties, and many more. Depending on the observation site also, other effects such as fog or water logging might occur. Local wind systems such as mountain-valley winds or sea-land winds, and Luv-Lee-Wind systems might also influence the local climate leading to cold-air flows and pools. As mentioned earlier also, the individual influences of high- and low vegetation and mountain tops need to be considered. Similar to urban climate studies, the appropriate placement always depends on the research questions to be answered.

2.6.4 Representativeness

Sensors record the environment to which they are exposed. Hence, **the turbulent and radiation source areas (footprints; see** ► Sect. 1.2.4) **need to be carefully considered in complex surroundings**. Sensor placement needs to consider wind directions, seasons, and local wind systems. For cities also the radiation source area can be very heterogeneous and needs to be carefully considered regarding the siting of the instruments to avoid unrepresentative radiation and shadowing effects on radiation and temperature variables. It is often advisable to look for a homogeneous patch representing the typical conditions in a specific local climate zone to avoid too much complexity and make the gathered data interpretable and useful. There are a few right or wrong regarding the placement of measuring instruments, as it depends on the scale, representative, and the respective research question. This means that attaching a sensor to a building with a distance of fewer than 0.5 m from a building wall may be suitable for studying the effect of the building itself, but it is rarely representative of the street canyon, neighbourhood, or city. Ideal observation sites for most urban research questions at street, neighbourhood, or city scale are within the urban canopy layer and well below the average roof height, in an as far as possible homogenous surrounding within several hundreds of meters while avoiding disturbance such as heat releases from buildings (◘ Fig. 2.9). When observing within a street canyon, the Height to Width Ratio ($\frac{H}{W}$) and surface and building properties should be representative of the LCZ (WMO 2023). In general, north–south-directed streets are favoured, because the temperature signal is less distorted, while the daily temperature amplitude might be more pronounced. In other cases to avoid wind effects and channeling, an east–west direction might be more suitable (WMO 2023). When setting up an urban meteorological network, also the site of the street needs to be controlled and measurements should be taken on the same sites (WMO 2023).

Also, vertical siting can be challenging in urban settings. General meteorological observation guidelines may need to be adjusted for urban studies because urban areas are highly diverse. Often available stands such as street lights or traffic signs are used for mounting air temperature sensors (Stewart and Mills 2021). This can help to uniform the measurement height and might grant access to power. Safety and vandalism can be concerns while finding suitable locations for placing sensors, leading to special criteria for sensor sites and heights. For instance, sensors for air temperature observation in urban areas are often placed higher than the standard 2 m to prevent vandalism, making them out of reach without the need for ladders.

In urban environments with diverse property owners, securing agreements, permissions, and, if necessary, contracts for sensor placement takes considerable time. Additionally, regulations for the protection of historical monuments and other factors may also need to be considered.

Urban usage has specific requirements such as the compactness and power supply of the sensors. These sensors are often placed on existing infrastructure like street lights, traffic signs, or roofs, which can impose limitations regarding weight and dimensions, necessitating the use of more compact sensors. If no power supply is available, solar panels are often used, and the sensors' dimensions and power supply must be constrained to minimize additional wind load on the existing infrastructure. Urban environ-

◘ **Fig. 2.9** Example for an unrepresentative measurement site (left) due to different LCZ within the radius of 100 m (outer ring) and a more representative station placement (right) within homogeneous environment and a potential representative footprint (► Sect. 1.2.4) source area (right)

ments may have more data transmission options due to the wider availability of Wi-Fi or LoRaWAN.

When collecting urban data documentation, certain metadata such as sensor height, LCZ, human activity and heat releases, and short-term changes such as constructions are particularly important (WMO 2023).

2.6.5 Guidelines and Recommendations for Urban Observations

The WMO (2023) provides some guidance for measuring the urban canopy layer. Before looking for a specific location, the following considerations need to be made (WMO 2023):

- "Is the station intended to characterize the greatest impact of the urban area on temperature (use cases such as weather forecasting or research)?
- Is the station targeting the area in the city with the highest population (use cases such as energy use or health impact)?
- Is the station intended to represent a block, a neighbourhood, or a larger urban area (use cases such as climate monitoring or urban design)?
- Is the station part of a meteorological network intended to define the intra-urban temperature variability for specific neighbourhoods and the associated spatial structure (use cases such as health impact or urban gardening applications)?
- Is the station targeting a particular site (for instance intended for development or redevelopment) or is it intended to characterize a particular environmental issue or issues (use cases such as ecology and vegetation or atmospheric chemistry)?
- Is the station intended to be in place for a short experiment or a long period (longitudinal study), potentially before and after heat mitigation interventions are put in place (use cases such as urban design and planning or energy use)?"

For sensor placement and location, the following aspects need to be considered (WMO 2023):

- When aiming to monitor local urban effects, it's crucial to assessing mesoscale influences such as cold air drainage, river valleys, and shoreline locations meticulously. Since local-scale cold air drainage effects can affect both urban and rural areas, it's essential to determine whether they should be accounted for within the site setting or not.
- Sensors ought to be positioned within a relatively uniform section of the Local Climate Zone (LCZ) and kept away from boundaries between LCZs exhibiting distinct surface characteristics. Sensors located within the Urban Climate Layer (UCL) are likely influenced by the surrounding environment within a radius extending up to 500 to 1,000 m (WMO 2023).
- Sites featuring unusual structures, surface covers, materials, or other properties within the thermal source area should be avoided. This is especially pertinent to areas with abnormally high or low moisture levels, or locations near concentrated heat sources such as heating plants or outlets of heating, ventilation, and air conditioning (HVAC) systems.
- While the standard screen-level height is 1.25–2.0 m above ground, this recommendation can be relaxed to allow greater heights in urban areas if the standard height is not possible (WMO 2023). The deviation might be small, due to the increased turbulence and vertical mixing in the urban environment. Consequently, data gathered at heights ranging from 3 to 5 m exhibit only minor discrepancies compared to measurements taken at the standard screen-level height (WMO 2023). However, rooftops should be avoided for most applications, as sensors placed there fail to capture canopy layer temperatures.

2.7 Specific Methods Applied

This section focuses on the different methods and tools used to collect meteorological data, emphasizing stationary, mobile, and

2

Table 2.6 Review of representativeness and measurement requirements for selected measurement methodologies

	Representativeness		Requirements	
	Spatial	Temporal	Instruments	Installation
Stationary measurements	Limited to installation location	High	High-quality, calibrated	Fixed location, no obstructions
Mobile measurements	High	For short term periods, depending on task of campaign	Portable or vehicle-mounted	Vehicles (car, bus), bicycle, tram, train, drone, ship, aircraft
Transect measurements	High, but limited to chosen path	According to the duration of the measurement	Portable	Manual or mobile deployment along a transect
Crowd-sourcing	Potentially high	Variable	Consumer-grade sensors, smartphones	Numerous volunteers contributing data

transect measurements, as well as the increasingly popular crowd-sourced measurements. Further, the importance of representativeness of meteorological measurements is discussed to ensure that the data accurately reflects the conditions of the broader area (Table 2.6). It covers also the technical aspects of instrument calibration, maintenance, and the challenges associated with different measurement techniques. The integration of stationary and mobile measurements provides a holistic approach to meteorological data collection, enhancing our ability to analyse and respond to atmospheric processes.

Despite awareness of its effects, it is sometimes impossible to avoid the impact of objects or energy sources on measurements. To minimize these effects and categorize "non-perfect" measurements, the WMO classification (WMO 2018/2024) is used. This guide specifies different classes of meteorological measurements based on instrument quality and accuracy while introducing its exposure and maintenance as important factors. These classes typically range from Class 1 (the highest standard) to Class 3 or lower. Descriptions of top three classes and their general criteria (accuracy, siting, maintenance, data quality) are as follows.

Class 1. Instruments of **highest** accuracy, **optimal** siting (ideal location with no obstructions; optimal exposure), **frequent** calibration and maintenance, **high** data quality (high accuracy and reliability).

Class 2. Instruments of **high** accuracy, **good** siting (good location, minimal obstructions, some minor environmental influence), **regular** calibration and maintenance, and high data quality (slightly lower than Class 1).

Class 3. Instruments of **moderate** accuracy, **fair** siting (acceptable location, moderate obstructions, significant environmental influence), **periodic** calibration and maintenance, **moderate** data quality (affected by environmental factors).

2.7.1 Stationary Measurements

Stationary measurements, also known as fixed or permanent measurements at weather stations, are based on devices permanently installed at a specific location to continuously measure and collect values of meteorological elements. These sensors are invaluable for long-term (climate) monitoring in both urban and rural areas. However, some tasks require measurements at only one or a few specified locations for a limited period of time, e.g. for analysing specific local weather or microclimatic phenomena. In these instances, "stationary" primarily refers to the fixed location and the time-averaged measurement at that site. The averaging time must exceed the device's specific time constant (τ). **Typical averaging times for micrometeorological measurements are 1, 10, 30, and 60 min**. If shorter averaging time is needed, such as 1 min, the instrument's response time is crucial. For longer averaging times, like 60 min and longer, the data already reflect the average daily variations of the meteorological variables.

Micrometeorological networks consist of a series of spatial distributed weather stations deployed in a manner to sample **microscale** features of local micrometeorological phenomena (▶ Sect. 1.2.2). These phenomena, ranging in size from tens to hundreds of meters, can evolve rapidly within minutes. Typically, all network sites share a common infrastructure, sensors set, and sampling methodology. Data from station networks are collected together, often with real time access, and after quality-control being archived and disseminated.

Network measurements represent a specialized branch of meteorological measurements, often referred to as "measurements for special purposes" for operational or research applications. The primary objective of a micrometeorological network is to collect similar variable observations simultaneously in time and space across a designated region to achieve this specific goal. Each network configuration must be meticulously designed to capture the expected microscale location features accurately.

Each site requires appropriate equipment like platforms, sensors, power, and communications, installed at standardized or process-relevant heights. A central processing centre is needed in order to collect, process, quality control, archive, and disseminate data to users. Comprehensive metadata should accompany the data, detailing site locations and characteristics, configuration, sensor properties, and quality control standards. For both research and operational purposes, these measurements must be accurate, precise, and continuous in real time. More details about network measurements can be found in (Foken 2021a, Chaps. 45, 63, and 64).

2.7.2 Mobile Measurements

The heterogeneity of terrestrial surfaces requires high spatially resolved measurements of atmospheric quantities in time and space. Since stationary measurements are only informative for a small spatial area in a complex terrain, mobile measurements guarantee a higher resolution at least along the measurement routes. However, the effort for such measurements is very high, so that one concentrates on single typical weather conditions, usually autochthonous weather conditions, which show a pronounced diurnal variation of air temperature, humidity, radiation, and partly also local wind systems under anticyclonic conditions. We distinguish between horizontal and vertical mobile measurements. A basic requirement for mobile measurements is the recording of the correct positioning and timing of the "moving" measurements points during a drive, which is usually determined by an included Global Navigation Satellite System (GNSS). Horizontally moving measurement systems are increasingly used in meso-meteorology and especially also in urban climatology (Parlow and Foken 2021; Foken 2021a, b; Foken et al. 2023; VDI 2023). These are mounted either on ground-based vehicles such as bicycles, car, bus, and tramway or others. An example of a bicycle-based measurement system, designed to investigate the air quality of Stuttgart (Germany) and the impact of local traffic on the pollutant concentrations, is shown in ◘ Fig. 2.10 (Samad and Vogt 2020).

An example result of a ground-based mobile transect measurement by bicycle is shown in ◘ Fig. 2.11. The measurement with bike was done on "Donaufeld", an agricultural landuse area surrounded by urban structure of the city of Vienna, Austria, and a Danube river side arm at the south side. The sensor was mounted at 1 m height on the front of the bike. So the sensor was well-ven-

2

Fig. 2.10 Mobile measurements using a bicycle platform named MOBAIR. From Samad and Vogt (2020), © Authors CC BY 4.0. *Note* The weather station must be orientated vertically for wind measurements

Fig. 2.11 Transect mobile measurement by bike of night air temperature (03.08.2017 at 3:10am) within an urban environment in Vienna, Austria. *Source* Mursch-Radlgruber

Table 2.7 Classification of mobile measurement systems (Parlow and Foken 2021)

Orientation of data collection	Scale	Typical systems
Vertical mobile systems	1–100 (1,000) m	Lifts on masts, balloons, UAS
Horizontal mobile systems	10–100 m	Systems on rails or ropes
	100–10,000 m	Cars, streetcars, bicycles, etc.
	100–10,000 m and larger	UAS

a)

b)

Fig. 2.12 UAS (drone) used for meteorological measurements: **a** Fixed-wing aircraft, © Moritz Mauz, University of Tübingen; **b** Multicopter, © Jens Bange, University of Tübingen

tilated and was radiation-shielded by a small climate hut. The example represents night-time cooling (start of the measurement at 3:10am till 4am) showing maximum temperature of 24 °C in the built-up area and approximately 20 °C in the agricultural area. So the maximum temperature cooling by the green area within the urban environment was about up to 4 °C.

Another sensor transport system are unscrewed aerial systems (UAS, called drone) (Table 2.7). Since atmospheric stability is of considerable importance, especially for UAS measurements, other vertical moving systems are also listed in Table 2.7. For more details on vertical UAS measurements, see ▶ Sect. 2.3.

Moving instrument carriers (Fig. 2.10) typically move continuously or make brief stops specifically for wind measurement. For UAS (Fig. 2.12), the type of device and the flight patterns determine the application (Table 2.8). Measurement of turbulent fluxes is largely limited to fixed-wing aircraft. It should also be noted that characteristics of the surface can only be detected at flight altitudes below the blending height, above which only area-averaged data are recorded at the landscape scale.

The time duration of mobile measurements should be chosen so short that no temporal change of the meteorological fields occurs during the measurements. This can hardly be realized for measurements on the ground, so that reference stations at fixed locations have to be included in the measurement cycle in order to be able to correct the temporal change of the measurement signals.

The sampling rate of the signals and the travel speed are decisive for the spatial resolution of the measurements (Table 2.9). As a rule, however, it is not the sampling rate that limits the spatial resolution of mobile measurements, as this is now usually 1 s due to modern electronics, but the time constant of the sensor is the limiting factor. The spatial resolution of a moving sensor with a time constant τ (Note 2.2) can be described by a path length A, which (in a first linear approximation) is related to the travel speed v, the

2

Table 2.8 Comparison of the measurement capabilities of fixed-wing aircraft and multicopters. For hybrids, the corresponding applications apply depending on the flight mode (VDI 2021): + suitable, (+) conditionally suitable, − not suitable at the current stage of development; 1: flow measurements with horizontal flight pattern; 2: static stability with vertical flight pattern; 3: dynamic stability with vertical flight pattern

Measured variable	Fixed-wing aircraft		Multicopter	
	Mean value	Turbulence	Mean value	Turbulence
Horizontal wind (3)	+	+	(+)	–
Vertical wind (1)	+	+	–	–
Air temperature (1–3)	+	+	+	–
Water vapour concentration (1)	+	+	+	–
Gas concentration (1)	+	(+)	+	–

time constant, and the adjustment factor f_τ according to Table 2.10, according to the following relation:

$$A = v \cdot f_\tau \cdot \tau. \tag{2.5}$$

For the practical application of Table 2.10, one estimates the possible difference of the temperature during the measurement run and determines the percentage share of the measurement error in the expected difference. The difference of the percentage of the measurement error to 100% is the percentage approximation to be assumed and thus the adjustment factor to be applied in Eq. (2.5).

The input signal of most sensors will follow a very sudden change in the measured quantity with certain delay and damping effects. These effects are commonly characterized by an individual sensor response time, or time constant (Hübner et al., 2014). In mobile measuring systems, the dynamic error leads to a spatial and temporal offset of the measured values (Fig. 2.13). In case of a linear temporal or spatial change of the measurement signal, a correction is possible (Note 2.2).

Note 2.2

Time constant and its correction

The temporal change of the quantity of a measurement signal can be described by the following differential equation (Brock and Richardson 2001; Parlow and Foken 2021; Foken and Mauder 2024):

$$X_s(t) = X_i(t) + \tau \frac{\mathrm{d}X_i}{\mathrm{d}t}. \tag{N2.1}$$

Here X_i is the input signal and X_s the output signal, influenced by the time constant τ (the inertia of the sensor), and t is the time. For the case of a linear change of the signal ($X_i(t) = a \cdot t, t \geq 0$) with time (analogously also in space) the change of the signal per time unit a follows

$$X_s(t) = a \cdot t - a \cdot \tau\left(1 - e^{-\frac{t}{\tau}}\right) \tag{2.2}$$

where the second term on the right-hand side of the equation is called the dynamic error (see Note 1.15), which describes the difference between the measured signal and the input signal at a given time:

$$\Delta X_d(t) = X_s(t) - X_i(t) = a \cdot \tau\left(1 - e^{-\frac{t}{\tau}}\right). \tag{2.3}$$

The output signal is shifted with respect to the input signal by the time difference $\Delta t = \tau$ (Fig. 2.13). Equation (2.2) can be used to correct the dynamic error.

The duration of a measurement run should only be selected long enough to ensure that the meteorological conditions change only insignificantly in order to avoid extensive trend correction. Table 2.11 gives hints for typical measurement runs of different mobile systems.

Table 2.9 Spatial resolution between two measuring points in meters of mobile measuring systems as a function of speed and sampling interval (Foken and Mauder 2024)

Sampling interval in s	Driving speed in km h^{-1}				
	5	10	20	30	50
1	1.4	2.8	5.6	8.3	13.9
2	2.8	5.6	11.1	16.7	27.8
5	6.9	13.9	27.8	41.7	69.4
10	13.9	27.8	55.6	83.3	138.9
20	27.8	55.6	111.1	166.7	283.3

Table 2.10 Required measurement duration for a percentage approximation to the real value with a step change in the signal as a multiple of the time constant (VDI 2023)

Percentage approximation	50%	63.2%	80%	90%	95%	99%	99.9%
Adjustment factor f_τ	0.7	1.0	1.6	2.3	3.0	4.6	6.9

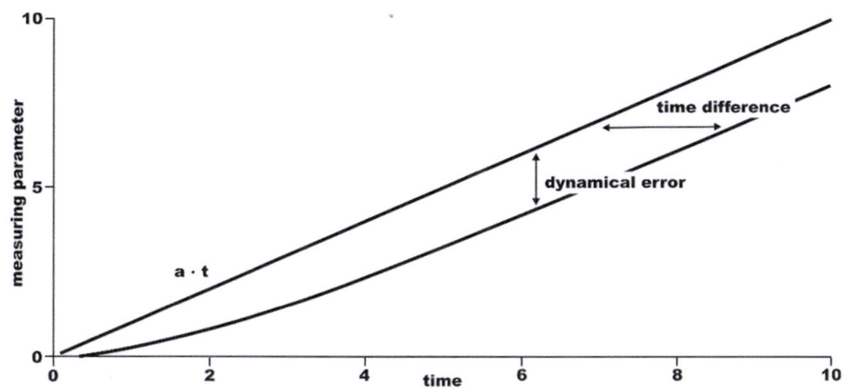

Fig. 2.13 Schematic course of the dynamic error and the time difference in case of a linear change of the input signal (a: change of the measurand per time unit; t: time) from Foken and Mauder (2024) with kind permission of Springer

Linear time correction is a simple procedure for correcting air temperature measurements, provided that they are subject to an approximately linear trend during the measurement period. Here, the temperature change between the start and end points of the measurements or between points approached twice in the case of large areas under investigation and is assumed to be linear. In the case of measurement sections, where the start and end points are spatially far apart, the correction must be made taking into account one or, if necessary, several reference stations. This applies in particular if areas of different land use show a clearly different cooling or warming behaviour. For temperature and humidity measurement, radiation protection is also absolutely necessary for mobile systems. Since this increases the bias of the sensor, the system must be force-ventilated if the ventilation is not sufficient due to the speed of travel.

Maintenance should be carried out as recommended for the applied sensors. Because of the high mechanical stress on mobile systems, maintenance should not only be per-

2

Table 2.11 Application ranges of different instrument carriers for a maximum measurement duration of 2.5 h (VDI 2023)

Instrument carrier	Typical area size in km^2	Typical speed in $km\ h^{-1}$	Recommended route length in km
Carrying frame, hand cart	0.1 × 0.1 to 2 × 5	2 to 4	about 5 to 10
Bicycle	1 × 1 to 5 × 10	10 to 20	about 25 to 50
Motor vehicle	2 × 5 to 10 × 20	25 to 40	about 50

Table 2.12 Comparison of various mobile systems with various fixed multi-sensor systems (Parlow and Foken 2021)

	Mobile systems	Fixed multisensory systems (stationary measurements)
General remarks	Can only be used for short measurement periods Often use simple or miniaturized sensors, a high level of quality control is required	Useful for short- and long-term measurements Application of recommended sensors (WMO 2024) is possible Possibility of time shifts between measurement points and dynamic error
Large-scale horizontal measurement systems	Larger areas are covered, high flexibility in terms of area studied (dependent on weather conditions)	Only a few measuring points, no flexibility in terms of area studied, small-scale horizontal measurement systems
Small-scale horizontal measurement systems	High spatial resolution, dynamic error is possible	Only a few points can be observed (see Transect Measurements)
Vertical measurement systems	Needs expensive instruments with a small offset and time shift	Highly accurate systems are necessary (tower measurements)

formed at predefined time intervals (just as for fixed sensors) but should also be adjusted according to the frequency of use of the mobile system. The instruments must be in working order and checked at the end of the measurement programme. New measurement programmes should be preceded by re-calibration. Reference measurements should be made with high-quality sensors, and the siting conditions should be of the highest class (WMO 2024).

Quality control is a very important issue for mobile systems due to the high mechanical stress these systems endure and their often low accuracy related to effects such as delay time and dynamic error. Comparison with reference instruments allows (i) correction for changes in the values of the measured meteorological variables during the time taken to perform the itinerary. The system should pass the reference measurement station as often as possible; (ii) re-calibration of simple sensors or sensors where there is time drift; (iii) examination of the measurement differences from different locations to determine whether the differences are due to natural causes, incipient breakdown, or contamination of the device.

Finally, a comparison of fixed and mobile systems can help to decide when to use one or another measurement system (Table 2.12). For mobile measurements, it is often advisable to stop (hover) at one or two locations and take the "stationary" measurement, preferably several times in succession, taking into account the inertia (τ) of the instruments.

2.7.3 Transect Measurements

Transects are used methodically in different scientific disciplines that establish a spatial reference to the object under investigation.

Here we focus on agrometeorological applications, where a transect is basically a set of measurement or observation points along a horizontal path from a micro- to mesoscale. It can be planned as a specific designed transect to investigate certain microclimatic phenomena as well as taken as a part of a suitable station network or applying "mobile measurement" using a single mobile sensor over a horizontal path. The measurement transects don't need to be necessarily linear, depending on the specific conditions such as orography or research question (see ▶ Sect. 2.7.2).

Specific methods which are not further outlined here are based on electromagnetic or acoustic technology, where over a horizontal beam with an emitter and a receiver integrated specific atmospheric conditions and processes can be detected (e.g. using laser for sensitive heat flux estimate by a Scintillometer; see ▶ Sect. 3.12). For further reading see Foken (2021a).

The method of single-point measurements over a horizontal path frequently is used in agrometeorological and urban applications. It is characterized by stations in the horizontal path of the same type and design, estimating single or several meteorological variables at the same height levels (single or multiple heights) above ground.

For example, in order to detect wind fields behind a hedgerow, a number of wind sensors (stations) can be placed in different distances from the hedgerow to detect the hedgerow wind breaking effects (◘ Fig. 2.14a). Changing wind speed with the distance from the hedgerow is influencing also other microclimatic variables, such as evaporation (◘ Fig. 2.14b), air temperature and humidity, precipitation, leaf wetness, and radiation (in combination with shading effects), which all can be measured at the same time as well.

Such transect measurements can be applied on a (horizontal) microscale for detecting microclimatic differences in inhomogeneous plant stands or canopies such as in Intercropping systems as well (◘ Fig. 2.15a, b) or at the mesoscale within the landscape or valleys (◘ Fig. 2.16a). For example, over hilly terrain local mesoscale climatic characteristics (such as temperature profiles or cold air accumulations) affected by the terrain can be spatially mapped and related to terrain characteristics (◘ Fig. 2.16b).

2.7.4 Data Acquisition of Measurement Systems

In this section, based on the introduction in ▶ Sect. 1.3, we address the technical aspects of meteorological measurements and data collection, from sensor types through the data collection process to data logging as-

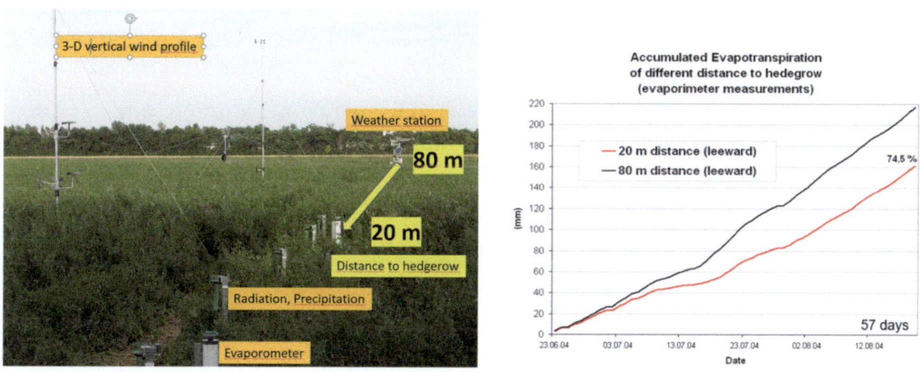

(a) (b)

◘ **Fig. 2.14** **a** Linear transect measurement setup for the wind field characteristics and other related micrometeorological variables leeward of a hedgerow; **b** evaporation measured by evaporometers in different distances from that hedgerow (BOKU, Marchfeld, Austria). *Source* Eitzinger, BOKU

(a) (b)

Fig. 2.15 **a–b** Transect measurement and detection of microclimatic variables, including evaporation (by automatic evaporometer) and the horizontal wind field (ultrasonic sensors) and other variables within two different strip intercropping field experiments (Project InterCropValues, BOKU, Vienna Austria). *Source* Eitzinger, BOKU

(a) (b)

Fig. 2.16 **a** Orography-oriented transect measurement of microclimatic variables (temperature, air humidity, and evapotranspiration) over a hilly vineyard terrain (larger scale), Göttlesbrunn, Austria; **b** Measured daily minimum temperatures (°C) over a transect of a vineyard terrain during an early frost event. *Source* Eitzinger, BOKU

pects. Every micrometeorological measurement begins with a sensor selected to measure the specific variable.

Sensors (see ▶ Chap. 3) provide (i) continuous-time (analogue) signals in response to the physical phenomenon, and (ii) act as transducers, converting physical stimuli into continuous electronical signals that reflect real-time changes in the environment. Modern sensors typically utilize various electronic principles, including resistance, capacitance, inductance, and/or voltage, to measure and represent variables. Analogue signals are continuous and can vary across a range of values, unlike digital signals, which are discrete and represented by binary values (0 and 1). The output of a sensor is typically a voltage (e.g. 0–1, 0–5, 0–10 V) or current signal (e.g. 4–20 mA) that changes in direct proportion to the variable being measured. For example, a temperature sensor may output a voltage signal that corresponds to the environmental temperature. The relationship between the measured quantity (sensor resistance changes due to heating or cooling, for example) and the output signal is often linear, although, in some cases, it can be nonlinear and requires additional conversion (see sensor calibration procedures below).

Calibrationof sensors is crucial to making measurements comparable and avoid measurement errors (see ► Sect. 4.2) or biases between sensors (which is important for all applications, and especially for measurements which require high accuracy such as measuring profiles or transects). **Sensor calibration factors** are provided normally by the selling companies in the technical descriptions. However, it is good practice to compare the set of sensors to be installed (new or especially older ones) of the same type before starting a measurement campaign. Sensors may need **re-calibration** after some years due to material ageing causing drifts in the measured signals (e.g. especially for radiation sensors and wind cup anemometers) and others.

Similarly, **maintenanceof the sensors** (see ► Chap. 3) is the basis for reliable measurements even for automatic systems. It is necessary to determine the frequency of sensor and station checks (daily, weekly, monthly, semi-yearly, etc.) and of what tasks should be performed. This includes the station's (i) remote control, (ii) the quality assurance of the measured data, and (iii) the elaboration of station visits (Sturtevant et al. 2021).

Sensors can transmit analogue signals over wires to the data acquisition device (datalogger), which therefore need to have wire connectors. Dataloggers digitalize the analogue signal using **analogue-to-digital converters (ADC, see below) or may receive already digitalized signal from a sensor device**. Digital-output sensor devices always require an external power supply to measure and transform the analogue signal into a digital one using the ADC included within the sensor device. These devices can then transmit the digital signal either via wired or wireless connections (e.g. Wi-Fi) to a datalogger or other receivers.

Single-ended versus differential-ended measurement or signalling is the simplest method of transmitting electronic analogue signals over wires. In this method, one wire carries a varying voltage that represents the signal, while the other wire is connected to a reference voltage, usually ground. For each signal to be transmitted, only one wire is needed, in addition to the common ground. Differential signalling, on the other hand, transmits (sends) the signal over two wires: one wire carries the main signal, and the other one carries the inverted signal (opposite polarity, equal magnitude). At the receiver (e.g. at the datalogger connectors), the magnitude of the difference between these two complementary signals is twice as large as the original signal. This allows the signal to be transmitted over longer distances. Signal lines (cables) that have the same length, resistance, and impedance are called balanced lines (or balanced signal pairs). Transmitting a signal using differential signalling over balanced lines gives very good resistance to noise and interference (see also Brock and Richardson (2001, Chap. 13 and Appendix C) and Foken (2021b)).

Sensor output calibration procedures have to be carried out before starting any measurements. After digitalization by an ADC in a datalogger or sensor device, the electronic output signal (e.g. mV) should be converted into the values of the specific meteorological variable unit (e.g. °C for temperature). For sensors with linear responses over the measurement range, a linear relationship between the sensor output versus the measurement range of a certain variable is given. Here, calibration factors such as a linear regression (see Note 2.3) are applied to translate the sensor output value (e.g. an electronic unit such as Volt) to the variable unit (such as °C) before or after storing the digital data in the datalogger storage. Typically, a linear regression approach (bias plus multiplier) is used for radiation sensors, which can differ even between the same sensor types, and are normally provided by the technical instructions. Standardized sensor types (such as thermocouples or resistance temperature sensors), as well as other sensors with nonlinear responses, often include the calibration calculation as part of a datalogger's sensor type settings or are incorporated directly into digital sensor devices, requiring no user interaction.

2

Note 2.3
Sensor output calibration—calculation procedure examples.
Two typical examples of often-used sensor types (for radiation and air humidity) are presented below.
Example 1: For sensors with linear response, linear regression in the form of

$$Y = A + B \cdot X \tag{N2.4}$$

is applied (where Y = value in the variable unit; A = bias value; B = multiplier value; X = sensor digital output value).
For example, on a radiation sensor following is indicated: 1 mV = 1 W m^{-2} (without a bias). The sensor digitalized output is 0.7 V. What is the value in W m^{-2} ?
(1V = 1000 mV; Y(W m^{-2}) = 1000(B) · 0.7(mV).
Result: The radiation sensor measures 700 W m^{-2}.
Example 2: In the case of a linear sensor with a fixed measurement range (such as an air humidity sensor), the measured value in the variable unit (%) can be calculated using the output value, the sensor range, and the output range as

$$Y = \left(\frac{(X - X_{min})}{(X_{max} - X_{min})} \cdot (S_{max} - S_{min}) \right) + S_{min} \tag{N2.5}$$

(where Y = value in the variable unit; X = sensor digital output value and its maxima (X_{max}) and minima (X_{min}); S_{max} = sensor maximum and S_{min} = sensor minimum in the variable unit).
Imagine a linear air humidity sensor having a range of 0–100% relative humidity (RH). Its analogue output is 16 mA in a 4–20 mA maximum range. What is the measured air humidity?
The sensor's measured value

$$Y = \left(\frac{16\text{ mA} - 4\text{ mA}}{20\text{ mA} - 4\text{ mA}} \cdot (100\text{ RH\%} - 0\text{ RH\%}) \right) + 0\text{ RH\%} = 75\text{ RH\%} \tag{N2.6}$$

Analogue-to-digital conversion(ADC) is the process of converting a continuous analogue (electronic) signal into a sequence of digital (binary) numbers. While the analogue signal is continuous over time, to convert this signal into a sequence of digital numbers, it has to be sampled and measured in regular intervals, defined by the sampling rate. The sampling rate should be fast enough to accurately track the signal changes as defined by the Nyquist-Shannon sampling theorem (Kester 2005). At each sampling moment, the continuous signal values should be measured, and approximated within a finite set of digital values, to obtain a digital number; this process is called quantization (see ◘ Fig. 2.17).

Both sampling and quantization introduce error or biases in ADC conversion. A faster sampling rate can better capture changes in the analogue signal, while increasing the resolution (number of bits) enables more accurate measurement of the signal's value (see Note 2.4). The connection and transmission of data between sensors, data acquisition systems, and datalogger and other devices is governed by standards and protocols that define physical (connector format and size), electronical (type and number of wires, voltage levels), and logical encodings (timing, number, sequence, meaning of bits). Some sensor and datalogger manufacturers use proprietary protocols, such as the Campbell Pakbus (2024) which are not further outlined here.

Note 2.4 Example: analogue to digital conversation
If the measured analogue signal voltage range is 0–5 V and the ADC sampling rate is 10 Hz, 8-bit resolution, then the analogue signal is sampled 10 times per second, each 0–5 V signal value is converted into an 8-bit number, and the 0–5 V range is converted to an 8 bit, 2^8 = 256 range from 0 to 255, resulting in a measurement resolution of 0.0195 V. Therefore, 0 V corresponds to the digital value 0, and 5 V corresponds to digital value 255. As the analogue range is

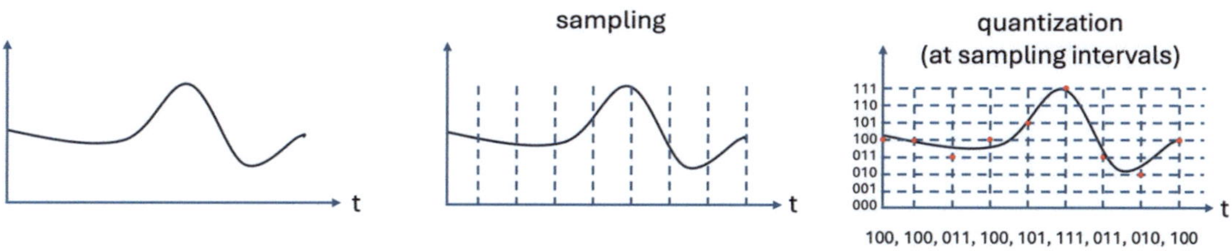

Fig. 2.17 Continuous analogue signal sampling and quantization. *Source* Zoltán Istenes, based on Kester (2005)

converted to the digital range, each digital value change corresponds to analogue value resolution (steps).

$$\frac{5\text{ V} - 0\text{ V}}{256} = 0.0195\text{ V} \tag{N2.7}$$

In professional applications, the choice between analogue sensors and digital sensor devices often depends on the type and requirements of the application, the availability of different types of sensors on the market, and their characteristics (analogue vs. digital, precision, accuracy). The decision is also affected by the capabilities of the data acquisition device (datalogger), the number, and the specifics of analogue and digital ports. Other factors, such as the number of sensors, the environment, and the required cable length, can also be decisive. **General advantages and disadvantages of both concepts** are listed in Table 2.13.

For building and design of own measurement systems, **dataloggers and basic data logging aspects** need to be understood. A datalogger receives the signals from analogue sensors and digitalizes it and/or receives digital values already from digital sensor devices and is a crucial part of a weather station. The datalogger can further process, store, and transmit the data following various protocols through different components (Table 2.14).

The sensor protocol, datalogger, and data transmission part of a meteorological station is often realized and implemented by a **microcontroller-based system** or by a single-board computer system in ready-to-use systems. Designing the functionalities of the microcontroller's program is a crucial aspect of creating an effective and efficient weather station. These are not further outlined here. Three examples of widely used datalogger types are shown in Fig. 2.18.

Table 2.13 Disadvantages and advantages of analogue versus digital signal transmission

Analogue sensor	Digital sensor device
Less electronics in sensors lead to general higher robustness against adverse weather conditions and lower data gap risks	Higher risks for failures in the sensitive electronic parts of sensors due to extreme temperatures, lightning, etc. Transmission problems can occur due to unsuitable wire connectors
Prone to signal noise, interference, longer cables increase inaccuracy (proper shielding and grounding required)	Digital transmission is more resilient to noise and interference, reserve the data (full precision) through the transmission
Limited signal processing, more time-consuming calibration and data processing	Unknown sensor calibration procedures may need user re-calibration or check
Requiring additional external components for advanced filtering or signal conditioning	Built-in ADC and circuit to create the digital format requires additional components and increases the price
Fast response and theoretically infinite resolution	The built-in ADC has limited sampling time, accuracy resolution

2

Table 2.14 Main datalogger components and aspects

Receiving sensor signals (interface or connector to sensors)	• Number and type of input ports
	o Analogue inputs
	▪ single ended
	▪ differential
	o Digital ports
	▪ Protocols (UART, CANBUS, RS232, I2C, etc.)
Processing, storing and transmission of sensor inputs	• Analogue ports, ADC conversion scan rate (frequency)
	• ADC conversion precision, resolution
	• Digital protocol parameters (max. baud rate, speed)
	• Sampling, statistical, and analytical functions
	• Data storage capacity (ring memory, non-volatile)
	• Data transmission (by wire or wireless)
Datalogger characteristics of relevance	• Environmental resistance (working temperature range, rugged, waterproof), weight, size
	• Level of programmability, remote control
	• Security, enclosure, on-site backup, display, control, lighting protection

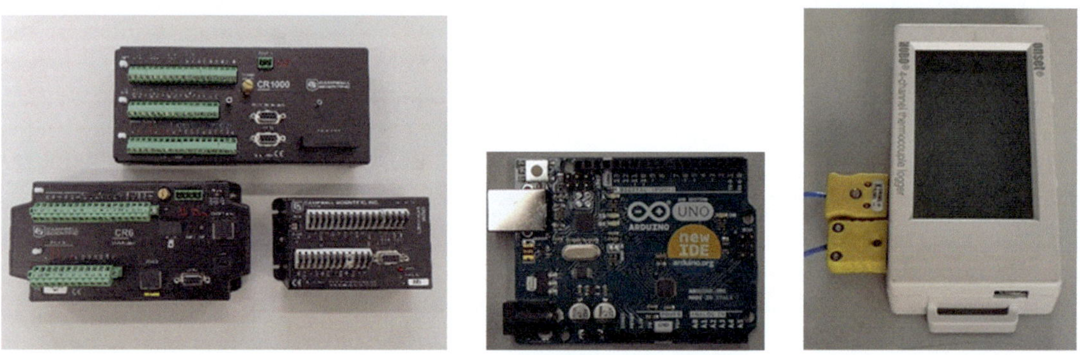

a) b) c)

Fig. 2.18 Examples of three different datalogger types: **a** Professional high-quality (research-grade, fully programmable) datalogger type Campbell; **b** Arduino Uno-based open-source microcontroller; **c** 4-channel thermocouple datalogger type HOBO—Onset with screen (all photos: Eitzinger, BOKU)

2.7.5 Design Principles of a Meteorological Station

The main goal of field measurements is to provide high-quality information and useable data (see more information from Pedhazur and Schmelkin 1991). A good measurement accurately and precisely captures the effects of processes in question with high reliability and minimal error. The main questions to address are (i) What meteorological quantities do we need to measure? (ii) What type of station and instrument configuration should we choose? (iii) Should we buy a complete station or build one from components? To help readers answer these questions, let's first review the main weather station categories.

Weather (or meteorological) stationscan be categorized using different criteria for its use and according to the field of application (◘ Table 2.15).

A set of **requirements and considerations for selecting or building weather stations or measurement systems** should be carefully analysed and checked before selecting, purchasing, or designing a suitable weather station, as summarized in ◘ Tables 2.16, 2.17 and 2.18. Commercially available weather stations typically offer standardized solutions with limited opportunities for customization, meaning that some specific requirements may not be applicable due to the station's predefined settings and characteristics.

2.8 Examples of Good Practices for Measurement Applications

To make the content of the previous sections a little easier to understand, this will be explained in more detail in the following sections using examples mainly from the field of agricultural meteorology.

2.8.1 How to Select a Representative Measurement Site—Agricultural Applications

A typical installation site in agriculture can be very different from a meteorological or hydrological application (◘ Fig. 2.20). Agrometeorological stations are normally installed

◘ **Table 2.15** Categories and characteristics of weather stations

Basic/Personal Weather Stations (◘ Fig. 2.19)	Typical low-cost designed stations for home use and the general public
	Often combined with indoor sensors and Wi-Fi connection
	Provide basic meteorological measurements like temperature, humidity, radiation, and wind speed (and more), often include low-cost or free (online) cloud service
	Quality and representativeness of measured data mainly depend on maintenance and siting conditions
Professional weather stations	More advanced, higher costs, suitable for professional measurements, e.g. urban and agro-meteorological applications
	Include a broader range of sensors for detailed meteorological measurements
Industrial weather stations	Designed for industrial applications
	Often include features like durability and resistance to harsh environmental conditions
Research-grade weather stations	High-precision and costs instruments used in scientific research
	Equipped with advanced sensors and data logging capabilities
Portable/handheld weather stations	Compact and easy to carry
	Suitable for on-the-go measurements in various locations
Automated weather observing systems (AWOS)	Typically used at airports for aviation purposes
	Provide real-time weather information for flight operations

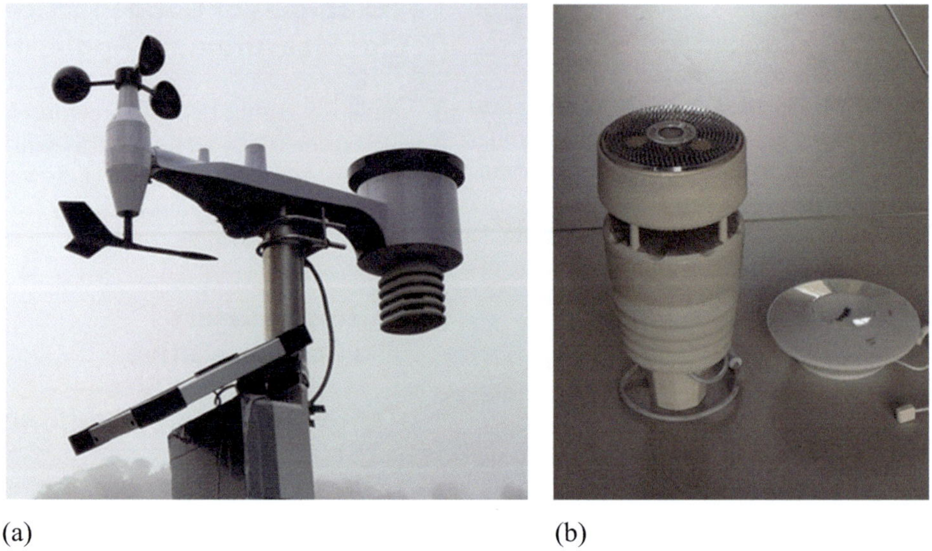

▪ **Fig. 2.19** Examples of online-access amateur weather stations in the 150–250 Euros category (incl. cloud service, temperature, relative humidity, solar radiation, precipitation, wind direction/speed, air pressure). Two examples of type Froggit/Ecowitt: **a** WS6006CE cellular wireless weather station (using sim card) and **b** GW2000A All-in-one sensor with Wi-Fi connection and external sensor options. Photo Eitzinger, BOKU

on the perimeters of a field (usually in the case of arable crops) or directly inside the crop canopy (e.g. annual crops or especially in perennial crops such as vineyards or orchards). Therefore, a weather station should occupy as little space as possible so as not to obstruct the daily crop management routine, such as spraying, cropping, pruning, and irrigating. As a result, the required space for an ideal installation is frequently unavailable, and a suitable compromise need to be found.

The selection of a representative site is especially important in mountainous areas or at locations with highly variable surface conditions (e.g. bare soil vs. crop field, as shown in ▪ Fig. 2.20) influencing strongly near-surface temperature spatial variability). In mountainous areas, flat land is scarce, with agricultural land alternating with forests, rivers, and lakes. All of these external factors can affect the weather station readings and form environmental conditions not fully representing, for example, a crop area. Furthermore, many stations are, for convenience, installed near farmhouses, which can obstruct wind flow and alter other measurements, creating microme-

▪ **Fig. 2.20** Photo and thermal image and histogram of a newly planted vineyard beside an established vineyard (Balatonfüred, 28 June 2022, 18:20 CEST (Central European Summer Time) (UTC+2); Photo: László Báder, BMGE, Hungary)

Table 2.16 a–c Checklist planning microclimatic measurement systems

(a) **Checklist for station's technical design and selection**
• **What should be measured?** Determine the specific meteorological variable which should be measured. This could include temperature, humidity, wind speed, wind direction, atmospheric pressure, precipitation, solar radiation, and more (in some cases also air pollutants). Consider that some variables, such as dew point, can only be measured indirectly
• **How should each meteorological variable be measured?** Choose the appropriate technology and sensor type for each variable according to measurement requirements and ability to handle the expected range of values in the settings environment. For example, wind speed can be measured using either mechanical or acoustic (ultrasonic) sensors, but if you want to register short term fluctuations or extremes, such as strong wind gusts, acoustic anemometers are the better choice
• **What is the required measurement range?** Define the expected range of values for each variable. Ensure that the selected sensors and measurement devices can handle anticipated conditions and expected range of values in the deployment environment. Consider that seasonal changes might affect the required frequency of data collection
• **What is the required precision and accuracy?** Specify the acceptable levels of precision and accuracy for each measurement. Different applications may have varying requirements, and the choice of sensors and instruments should align with these standards
• **How frequently should measurements be taken?** Specify the measurement frequency for each variable. Selected frequency should balance the dynamics of weather conditions, data temporal resolution, and the need for power-efficient operations
• **How should measurements be processed?** Determine the data processing algorithms and methods to convert raw sensor data into meaningful meteorological information. This should include planning for calibration procedures and addressing any potential sources of error
• **How should data be logged?** Specify the data logging mechanism, including storage capacity, logging intervals, data format, precision, and data retrieval methods. Consider the trade-offs between data resolution and storage requirements
• **How should data be transmitted?** Define the data transmission protocols, communication technologies (e.g. Internet, GSM, satellite), transmission frequencies, and data format. Consider bandwidth limitations, power consumption, and real-time data requirements
• **Should the data be aggregated?** Decide whether raw data should be aggregated or stored in its raw form (e.g. calculating and storing daily values instead of 10-min time interval values). Consider the trade-offs between data volume, processing complexity, and the specific needs of data analysis
• **Where will the measurement be conducted?** Specify the physical location of the weather station deployment. Consideration should be given to geographical and environmental conditions, including weather, impacts from electronic infrastructures (magnetic field ingress protection (IP)), and enclosures on sensor performance
• **Is there connectivity and power at the site?** Determine the required power availability (electronic network, battery, solar, etc.) and network connectivity (Internet, GSM, satellite) for the weather station
• **Sensor placement and position conditions?** Define the optimal placement and orientation of the sensors to avoid obstructions and ensure accurate measurements. This includes deciding on the optimal sensor height, avoiding exposure to direct sunlight, and other factors that could influence the data reliability

Table 2.17 a–c Checklist planning microclimatic measurement systems

(b) **Checklist on economic aspects for station design and selection**
• **Total costs occurring** Clearly define the budget constraints for designing, building, and maintaining the custom-built weather station. This includes costs associated with sensors, dataloggers, communication modules, power systems, enclosures, and any additional components or accessories required. The eventual subscription and/or fees based on the amount of transmitted data for the data transmission should be considered
• **Cost-effectiveness** Evaluate the cost-effectiveness of different sensor technologies and measurement methods. Consider the balance between upfront costs and long-term maintenance expenses to ensure a sustainable solution within the allocated budget
• **Maintenance and calibration costs** Anticipate ongoing maintenance and calibration costs that should include provisions for periodic sensor calibration, equipment checks, and potential replacements over the system's operational life
• **Return on investment (ROI)**
Evaluate the expected return on investment based on the value derived from the weather station data for professional or operational applications. It should be considered how the information gathered will contribute to the goals of the project or application and assess the long-term benefits relative to the initial investment
• **Vendor and component selection** Compare sensors, dataloggers, and other components prices coming from different manufacturers. Cost-effective solutions should be reached without compromising the required specifications and quality standards

teorological conditions that differ from those within the canopy of a certain crop.

A key benefit of a weather station installed in the plant canopy is its ability to generate time series of weather data that are site- and microclimate-specific. To do this effectively, the original conditions of the site at the time of first installation must be identifiable. It is therefore highly recommended to **take photographs of the installation** at a distance of about 20 m away, from all four directions (see ▶ Sect. 4.4.2 on meta information). When taking pictures, it is also advisable to record the GNSS coordinates of the weather station not only to be able to link the data to an exact location, but also to allow such mundane tasks as finding the equipment again when, for example, covered by a crop canopy.

Often agrometeorological weather stations have to be placed within crop fields or orchards to measure accurately microclimatic conditions (e.g. for disease monitoring, irrigation scheduling, or other specific applications). In such cases, the **changes in the growing canopy over time should be documented and reported** as well (through regular measurements of leaf area index, canopy height, and through notes or photographs) to support accurate data interpretation. For example, changes in canopy height and density during the measurement will significantly affect vertical gradients of micrometeorological variables.

2.8.2 Examples of Costs and Agricultural Applications

Depending on the measurement tasks, the required number of sensors, accuracy, reliability, sampling and logging options, and lifetime criteria determine the cost of the instruments, stations, or local measurement network and can vary widely (Table 2.19). Important factors to consider include available human resources, the frequency and cost of calibrations and maintenance, and the cost of telecommunications and amortization (see Checklists, Tables 2.16, 2.17 and 2.18). Emerging technologies such as new low-cost techniques for data transfer (e.g. LoRaWan) may reduce outdoor online station costs further in the future.

For many applications in agriculture, measurement networks or even station components are locally designed and built with

Table 2.18 **a–c** Checklist planning microclimatic measurement systems

(c) **Checklist for professional weather station applications: regulatory, compatibility, and others**
• **Measurement area-related requirements** Identify any unique or specialized requirements specific to the application area such as possible flooding, frequent walking line (vandalism risk), high slope with possible erosion, or presence of animals
• **Compliance with standards and regulations** Verify that the weather station design adheres to relevant meteorological standards and regulations, including requirements for accuracy, calibration, and data quality as per industry or regional guidelines
• **Compatibility with existing measurement systems** Evaluate the compatibility of custom-built weather station with existing devices, data management systems, and communication protocols to ensure smooth integration with current technology
• **Scalability, Future Expansion, and Modularity** Design the weather station with scalability, future expansion, and modularity in mind, ensuring that additional sensors or functionalities can be easily incorporated in the future. Modular components should be easily upgradable or replaceable
• **Environmental sustainability** Minimize the weather station's environmental impact by considering factors such as power consumption, materials used for components, and disposal at the end of its life cycle
• **Security and data privacy** Address data security and privacy concerns, especially if the weather station collects sensitive information. Implement encryption protocols, access controls, and data anonymization to ensure compliance
• **Emergency response capabilities** Incorporate features to support emergency response efforts, such as real-time data collection for early warning systems in disaster-prone areas. Ensure proper grounding and lightning protection where appropriate
• **Redundancy and reliability** Plan for redundancy to ensure continuous operation in critical applications. Include backup power systems or redundant sensors to minimize downtime and prevent data losses
• **Data visualization and user interface** Design or choose a user-friendly interface for easy data interpretation by diverse stakeholders, researchers, policymakers, or the general public
• **Accessibility and inclusivity** Ensure that the weather station is accessible to all target users. Consider display readability, compatibility with assistive technologies, and other diverse user needs

technical support, which further may reduce costs. Examples are described as follows.

Example 2.1 McCauley et al. (2021) developed a low-cost and open-source weather station (LOCOS) specifically designed to inform farm management decisions in a commercial agricultural operation (Fig. 2.21). The data collection and communication module could be developed for less than 200 €, but the total cost depends on the instrumentation. Integrated multi-sensor devices of high quality measuring several meteorological elements together can cost more than 2000 €, but could be of lower cost than single high-quality sensors.

Example 2.2 A similar further example of a modular station including the communication and data acquisition unit is illustrated in Fig. 2.22. The advantage of such farm-scale upgrades is that the stations can be easily customized/adapted in order to take in account financial constraints. The disadvantage is that own developments or technical adaptations may reveal unexpected defects during ongoing operation. To reduce relevant failure risks, a close, trusting relationship with the developer (e.g. a local company) of the measurement system is recommended. Alternatively modular systems can be built by fitting components from just one company.

Table 2.19 Ranges of costs of various options for low-to-medium cost measurement systems (stations) based on 2024 market cost estimates

Description	Cost range (2024) (€)	Application potential
Single sensors/handheld including data collection	50–500	Ad hoc measurements, short-term measurements
Simple low-cost outdoor stations, including all standard weather parameters and cloud service (see Fig. 2.19)	150–500	Suitable for basic weather observation, biases in measurements without further calibration possible; limited sampling periods; limited data transfer options (e.g. Wi-Fi) or sensor extension options
Agrometeorological station with minimum standard parameters, designed for field conditions (e.g. robustness in agricultural environment)	>1000	Suitable for weather risk observations (i.e. frost); limited sampling period options. Costs strongly depending on required type and number of sensors according to the application
Sophisticated, robust agrometeorological stations designed for field conditions, full set of parameters including profile measurements in air and soil, including data logging and data transfer service	2000–5000	Suitable for precision farming applications, irrigation planning. Model applications (e.g. pest warning systems), which need higher measurement accuracy
Mobile logging system, including GNSS	100–1000	Urban applications, temperature transects, etc.

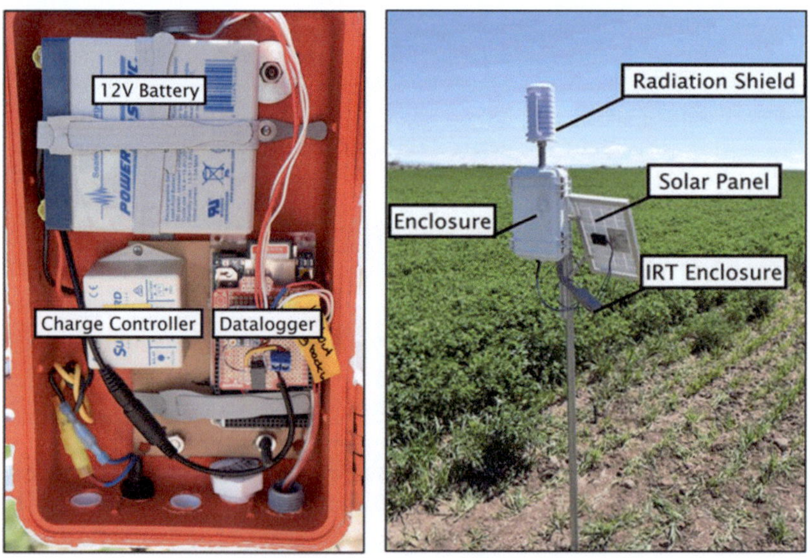

Fig. 2.21 Example of a low-cost agrometeorological station (internal components (left) and station external components (right)), from McCauley et al. (2021) with kind permission of Elsevier

Example 2.3 Since 1992 the Vintners Association of the Lower Austrian region has systematically built a network of >55 agrometeorological weather stations to feed a decision support system running various disease models, e.g. for downy and powdery mildew. The stations cover an area of well over 1.500 ha of predominantly white wine grapes. Data hosting, collection, and processing are done by the local office of Austria Chamber of Agriculture and Vintage School Krems (▶ https://www.vitimeteo.at), who provides

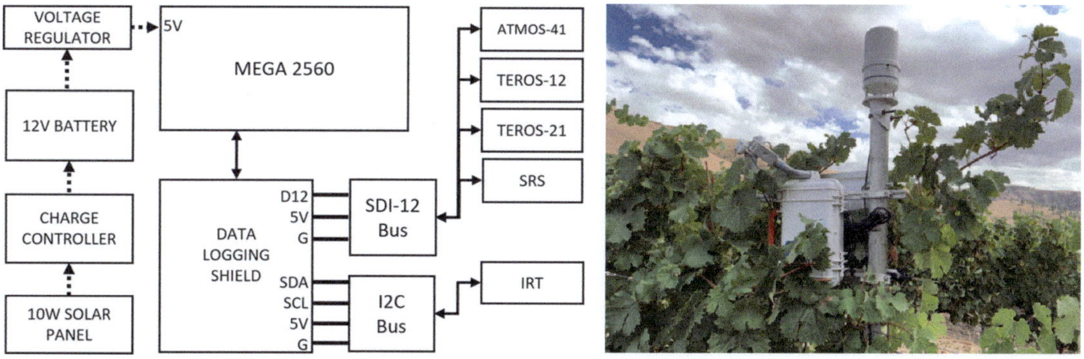

Fig. 2.22 Block diagram of the embedded computer system (left) and sensors used for a vineyard disease study (right). Dotted arrows indicate the power system; from McCauley et al. (2021) with kind permission of Elsevier

plant protection warning service through the Internet (▸ https://rebschutzdienst.at/) and other means. Members pay a very low annual fee for this service. Even without the extremely low cost for the service due to the ongoing support by the Chamber of Agriculture, the purchase of the equipment would have quickly paid back, as even a small 5 ha farm could achieve an annual saving of approximately Euros 600 and 200 Euros per spray, thus paying back a single station (approx. EUR 3000) in about 5 years. However, covering costs for long-term maintenance, data, and quality control remains a challenge if farmers expect organizational support.

- **Example 2.4** Forecasting and Reporting Service for Plant Protection of AP Vojvodina (PIS; ▸ https://pissrbija.rs/pocetna) conducts daily monitoring of the 150 harmful organisms in production of the 33 different crops and plant species. Regional PIS offices have been established in local agricultural stations, employing plant protection experts who are in charge of monitoring activities. 34 regional offices have been put into force across the country: 12 in Vojvodina and 22 in central Serbia. The scope of work also includes the monitoring of micrometeorological conditions in which the production takes place. Automatic weather stations (AWSs) located in crop fields and plantations are used to monitor the local weather as well as canopy microclimatic conditions. The most important meteorological elements, such as air temperature, soil temperature, relative air humidity, precipitation, and leaf wetness are monitored by 186 automated weather stations installed across Serbia.

References

Ács F, Horváth Á, Breuer H, Rubel F (2010) Effect of soil hydraulic parameters on the local convective precipitation. Meteorol Z 19:143–153. ▸ https://doi.org/10.1127/0941-2948/2010/0435

Ács F, Gyöngyösi AZ, Breuer H, Horváth Á, Mona T, Rajkai K (2014) Sensitivity of WRF-simulated planetary boundary layer height to land cover and soil changes. Meteorol Z 23:279–293. ▸ https://doi.org/10.1127/0941-2948/2014/0544

Barry R, Blanken P (2016) Microclimate and local climate. Cambridge University Press, Cambridge. ▸ https://doi.org/10.1017/CBO9781316535981

Baumgartner A (1956) Untersuchungen über den Wärme- und Wasserhaushalt eines jungen Waldes. Ber Dt Wetterdienstes 28:53 pp.

Beyrich F, Leps J-P, Mauder M, Bange U, Foken T, Huneke S, Lohse H, Lüdi A, Meijninger WML, Mironov D, Weisensee U, Zittel P (2006) Area-averaged surface fluxes over the LITFASS region on eddy-covariance measurements. Boundary-Layer Meteorol 121:33–65. ▸ https://doi.org/10.1007/s10546-006-9052-x

Biermann T, Babel W, Ma W, Chen X, Thiem E, Ma Y, Foken T (2014) Turbulent flux observations and modelling over a shallow lake and a wet grassland in the Nam Co basin, Tibetan Plateau. Theor Appl Climatol 116:301–316. ▸ https://doi.org/10.1007/s00704-013-0953-6

Breuer H, Ács F, Laza B, Horváth Á, Matyasovszky I, Rajkai K (2012) Sensitivity of MM5-simulated planetary boundary layer height to soil dataset: comparison of soil and atmospheric effects. Theor Appl Climatol 109(3):577–590. ▶ https://doi.org/10.1007/s00704-012-0597-y

Brock FV, Richardson SJ (2001) Meteorological measurement systems. Oxford University Press, New York. ▶ https://doi.org/10.1093/oso/9780195134513.001.0001

Cuxart J, Wrenger B, Martínez-Villagrasa D, Reuder J, Jonassen MO, Jiménez MA, Lothon M, Lohou F, Hartogensis O, Dünnermann J, Conangla L, Garai A (2016) Estimation of the advection effects induced by surface heterogeneities in the surface energy budget. Atmos Chem Phys 16(14):9489–9504. ▶ https://doi.org/10.5194/acp-16-9489-2016

Denmead DT, Bradley EF (1985) Flux-gradient relationships in a forest canopy. In: Hutchison BA, Hicks BB (Hrsg) The forest-atmosphere interaction. D. Reidel Publ. Comp., Dordrecht, Boston, London, pp S421–S442. ▶ https://doi.org/10.1007/978-94-009-5305-5_27

Finnigan J (2000) Turbulence in plant canopies. Ann Rev Fluid Mech 32:519–571. ▶ https://doi.org/10.1146/annurev.fluid.32.1.519

Foken T, Bange J (2021) Wind sensors. In: Foken T (Ed) Springer handbook of atmospheric measurements. Springer, Cham, pp S243–S272. ▶ https://doi.org/10.1007/978-3-030-52171-4_9

Foken T (Ed) (2021a) Springer handbook of atmospheric measurements. Springer, Cham. ▶ https://doi.org/10.1007/978-3-030-52171-4

Foken W (2021b) Principles of measurements. In: Foken T (Ed) Springer handbook of atmospheric measurements. Springer, Cham, pp S33–S47. ▶ https://doi.org/10.1007/978-3-030-52171-4_2

Foken T, Parlow E, Gross G, Leitl B (2023) Untersuchungsmethoden zum Stadtklima. Promet 106:97–114. ▶ https://doi.org/10.5676/DWD_pub/promet_106_10

Foken T, Mauder M (2024) Micrometeorology, 3rd edn Springer, Cham. ▶ https://doi.org/10.1007/978-3-031-47526-9

Garratt JR (1978) Flux profile relations above tall vegetation. Quart J Roy Meteorol Soc 104:199–211. ▶ https://doi.org/10.1002/qj.49710443915

Garratt JR (1980) Surface influence upon vertical profiles in the atmospheric near surface layer. Quart J Roy Meteorol Soc 106:803–819. ▶ https://doi.org/10.1002/qj.49710645011

Göndöcs J, Breuer H, Horváth Á, Ács F, Rajkai K (2015) Numerical study of the effect of soil texture and land use distribution on the convective precipitation. Hung Geogr Bull 64 (1):3–15. ▶ https://doi.org/10.15201/hungeobull.64.1.1

Holmes P, Lumley JL, Berkooz G (1996) Turbulence, coherent structures, dynamical systems and symmetry. Cambridge University Press, Cambridge

Hübner J, Olesch J, Falke H, Meixner FX, Foken T (2014) A horizontal mobile measuring system for atmospheric quantities. Atmos Meas Techn 7(9):2967–2980. ▶ https://doi.org/10.5194/amt-7-2967-2014

Kester W (ed.) (2005) The data conversion handbook. Newnes, London

Kolle O, Kalthoff N, Kottmeier C, Munger JW (2021) Ground based platforms. In: Foken T (Ed) Springer handbook of atmospheric measurements. Springer Nature, Cham, pp S155–S182. ▶ https://doi.org/10.1007/978-3-030-52171-4_6

Lalic B, Eitzinger J, Marta AD, Sremac AF, Orlandini S, Pacher B (2018) Agricultural meteorology and climatology. Firenze University Press, Firenze

McCauley DM, Nackley LL, Kelley J (2021) Demonstration of a low-cost and open-source platform for on-farm monitoring and decision support. Comput Electron Agric 187:106284. ▶ https://doi.org/10.1016/j.compag.2021.106284

Monson R, Baldocchi D (2014) Terrestrial biosphere-atmosphere fluxes. Cambridge University Press, New York. ▶ https://doi.org/10.1017/CBO9781139629218

Monteith JL, Unsworth MH (2014) Principles of environmental physics, 4th edn. Academic Press, Oxford. ▶ https://doi.org/10.1016/C2010-0-66393-0

Oke TR, Mills G, Christen A, Voogt JA (2017) Urban climates. Cambridge University Press, Cambridge. ▶ https://doi.org/10.1017/9781139016476

Panin GN, Foken T (2005) Air–sea interaction including a shallow and coastal zone. J Atmos Ocean Sci 10:289–305. ▶ https://doi.org/10.1080/17417530600787227

Parlow E, Foken T (2021) Ground-based mobile measurement systems. In: Foken T (Ed) Springer handbook of atmospheric measurements. Springer, Cham, pp 1351–1367. ▶ https://doi.org/10.1007/978-3-030-52171-4_50

Pedhazur EJ, Schmelkin LP (1991) Measurement, design, and analysis: an integrated approach. Psychology Press, New York. ▶ https://doi.org/10.4324/9780203726389

Peltola O, Lapo K, Thomas CK (2021) A physics-based universal indicator for vertical decoupling and mixing across canopies architectures and dynamic stabilities. Geophys Res Lett 48(5):e2020GL091615. ▶ https://doi.org/10.1029/2020GL091615

Samad A, Vogt U (2020) Investigation of urban air quality by performing mobile measurements using a bicycle (MOBAIR). Urban Climate 33:100650. ▶ https://doi.org/10.1016/j.uclim.2020.100650

Schönwiese C-D (2024) Klimatologie. 6. Aufl. Ed. Ulmer, Stuttgart

Shaw RH (1977) Secondary wind speed maxima inside plant canopies. J Appl Meteorol 16:514–521. ▶ https://doi.org/10.1175/1520-0450(1977)016%3c0514:SWSMIP%3e2.0.CO;2

Stewart ID, Oke TR (2012) Local climate zones for urban temperature studies. Bull Am Meteorol Soc 93(12):1879–1900. ▶ https://doi.org/10.1175/BAMS-D-11-00019.1

Stewart ID, Mills G (2021) The urban heat Island. Elsevier, Amsterdam. ▶ https://doi.org/10.1016/C2017-0-02872-0

Sturtevant C, Metzger S, Nehr S, Foken T (2021) Quality assurance and control. In: Foken T (Ed) Springer handbook of atmospheric measurements. Springer Nature, Cham, pp 49–92. ▶ https://doi.org/10.1007/978-3-030-52171-4_3

Thom AS, Stewart JB, Oliver HR, Gash JHC (1975) Comparison of aerodynamic and energy budget estimates of fluxes over a pine forest. Quart J Roy Meteorol Soc 101:93–105. ► https://doi.org/10.1002/qj.49710142708

Thomas C, Foken T (2007) Flux contribution of coherent structures and its implications for the exchange of energy and matter in a tall spruce canopy. Boundary-Layer Meteorol 123:317–337. ► https://doi.org/10.1007/s10546-006-9144-7

VDI (2021) Umweltmeteorologie - Meteorologische Messungen - Meteorologische Messungen mit unbemannten Flugsystemen (UAV) (Environmental meteorology—Meteorological measurements—Meteorological measurements with unmanned aerial vehicles (UAV)), VDI 3786 Blatt(Part) 22. Beuth-Verlag, Berlin

VDI (2023) Umweltmeteorologie - Methoden bodengebundener Stadt- und Standortklimamessungen mit mobilen Messsystemen (Environmental meteorology—Methods of urban and site-related ground-based climate measurements with mobile measurement systems), VDI 3785 Blatt 2 (Part 2). Beuth Verlag, Berlin

Venditti JG, Best JL, Church M, Hardy RJ (2013) Coherent flow structures at earth's surface. Wiley, Chichester

WMO (2024) Guide to instruments and methods of observation, WMO-No. 8, Volume I—Measurement of meteorological variables. World Meteorological Organization, Geneva

WMO (2023) Guidance to measuring, modelling and monitoring the canopy layer urban heat Island (CL-UHI), WMO No. 1292. World Meteorological Organization, Geneva

Yi C (2008) Momentum transfer within canopies. J Appl Meteorol Climatol 47:262–275. ► https://doi.org/10.1175/2007JAMC1667.1

Good Practices for Single Parameters

Thomas Foken, Branislava Lalic, Josef Eitzinger, and Tamás Weidinger

B. Lalic et al. (eds.), *Micrometeorological Measurements*,
Springer Textbooks in Earth Sciences, Geography and Environment,
https://doi.org/10.1007/978-3-032-03884-5_3

This chapter presents methods for measuring the individual meteorological elements. In contrast to other measurement technology books, the book has a completely different structure to help users quickly find the right method for them from the multitude of methods. With the help of notes and tables, the user can tick the methods and notes relevant to him in order to get a quick overview. First, the user is asked for the purpose for which the measurement is to be carried out (labelled with small letters). In additional tables, the user can then find out what accuracy is required for the measurement and how the measuring device is to be installed. The user is also informed of the meteorological processes and phenomena to be expected and is referred to the relevant sections in ▸ Chap. 2. Finally, the measuring devices are presented. The user is either referred to a specific measuring device according to the application or can select a device according to the specification required for the application. The selection of measuring devices was limited to commonly used ones. A complete overview of possible measurement devices can be found in the relevant literature (Foken 2021).

3.1 Air Temperature

Air temperature is a physical state variable and the most important meteorological parameter for describing weather, climate, and climate changes. The first quantifiable measurements of temperature date back to Galileo Galilei (1564–1642) and are dated 1592/93. Today, science uses the thermometer scale after Anders Celsius (1701–1744) in degrees Celsius (°C) and the thermodynamic temperature scale after William Thomson (1824–1907; Lord Kelvin of Largs) in kelvin (K). Since temperature is defined by basic physical quantities, it is repeatedly adapted to their accuracy, with the International Temperature Scale of 1990 (ITS-90) applying today (Preston-Thomas 1990).

Temperature measurements are common in all technical and scientific fields (Hebra 2010) and led to a variety of measurement methods and sensors. In the atmospheric field, there are certain limitations due to the need to adapt to the medium air, and in recent decades the increasing automation of measurements has had a limiting effect. In the following, we will further restrict ourselves to widely used and easily accessible measurement methods; further details can be found in the literature (Foken and Bange 2021a).

3.1.1 Why Should Air Temperature Be Measured?

Temperature measurements are needed for various purposes in meteorology and climatology. The application determines which measuring accuracy and which time constant the sensor must have. Typical application examples are given below, and, in the box, you can note the relevance for your own measurements (◘ Table 3.1).

◘ Table 3.2 contains the necessary accuracies and time constants for the applications. However, the measurement of air temperature is not trivial. If the probe is exposed to shortwave solar radiation, it heats up and becomes a radiation meter (bolometer). The error is called radiation error. Therefore, the sensor must be protected by a radiation screen or Stevenson temperature screen. However, this falsifies the measurements up to 1–2 K(hut error) and the time constant is several minutes. The hut error is caused by insufficient air exchange with the environment and still existing radiation error. A double ventilated radiation shield is more effective, and the ventilation rate (at the probe and between the two radiation shields) should be >2.5 m s^{-1}. The time constant can be reduced by ventilation, which is particularly important for application (f). In application (h), the air temperature is determined from the temperature dependence of the sound velocity, so that neither a screen nor ventilation is necessary. ◘ Figure 3.1 shows a typical radiation screen and a ventilated thermometer with double radiation protection.

Table 3.1 Application purpose for temperature measurements

Code	Application	Relevance
a	Input variable for weather, climate, and other models	
b	Climate analysis, local climate	
c	Characterization of human well-being → UTCI (Bröde et al. 2013)	
d	Determination of temperature-dependent physical quantities → e.g. air density, water vapour saturation, specific heats (Foken et al. 2021)	
e	Determination of water vapour saturation for measurements of air humidity → saturation water vapour pressure (Foken et al. 2021)	
f	Temperature measurements, e.g. with mobile systems for the study of the local climate → mobile measurements	
g	Determination of sensible heat flux from vertical temperature gradient → see gradient measurements (There are restrictions on the choice of location and measuring equipment!)	
h	Determination of sensible heat flux with the → eddy-covariance method (The eddy-covariance method uses → sonic anemometers to measure temperature. These measure the so-called sonic temperature, which corresponds approximately to the virtual temperature; see Note 3.4.)	

Remark: After selecting the application, you will find the necessary technical data and the sensor recommendation in the following tables; sign your application

Table 3.2 Accuracy and time constant for different applications (Foken and Bange 2021a) and need for a screen, if possible ventilated

Application	Accuracy (K)	Time constant (s)	Screen	Relevance
a, b	0.1	10–30	Yes	
c	0.2	10–30	Yes	
d, e	0.5	10–60	Yes	
f, g	0.1	5–10	Yes, ventilated	
h	0.05	<0.01	No	

3.1.2 Where to Measure?

Since the air temperature near the surface shows considerable gradients, the determination of a certain measuring height is necessary because of the comparability. The standard measuring height is 2 m (ISO 2015; WMO 2024). Table 3.3 contains information on the applications for which the standard measuring height is necessary or which deviating specifications apply.

The temperature measurement is also influenced by its environment. Therefore, special environmental conditions must be realized for applications a–c, or errors are possible if they are not given (Table 3.4; ISO 2015; WMO 2024). For applications g and h, see the corresponding measurement method (▶ Sect. 3.12.3).

For measurements that are not carried out over low vegetation as in applications a–c, special specifications apply with regard to the measurement height and possible processes that may influence the measurements. Table 3.3 is applied in the analogous way, but the influencing problems listed in Table 2.1 and the atmospheric processes listed in Table 2.2 must be taken into account.

3

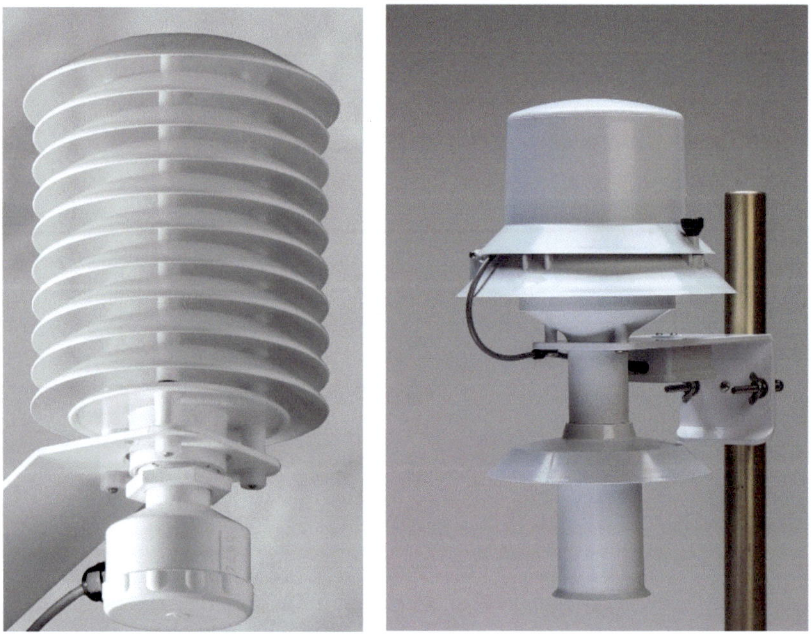

a) b)

Fig. 3.1 Nonventilated (**a**) and ventilated (**b**) temperature screens: **a** screen developed by Gill (model 41003); **b** ventilated temperature screen (model 43502). Photos © R. M. Young/GWU Umwelttechnik GmbH, Erftstadt, Germany

Table 3.3 Measurement height for different applications

Application	Measurement height	Relevance
a, b, c	2 m, recommended	
d, e	Height for which temperature measurement is required	
f	About 1–3 m	
g	Different heights between 0.25 and 10 m, logarithmical spacing	
h	Height of the sonic anemometer	

Table 3.4 Additional siting conditions for applications a–c

Class	Description	Error	Relevance
1	Flat area, covered with low vegetation (<10 cm), distance to heat sources or water >100 m		
2	Like class 1, but distance >30 m		
3	Like class 1, but vegetation height <25 cm and distance >10 m	±1 K	
4	Artificial heat sources and reflective surfaces less than 50% of the surface within 10 m radius, no shade when sun is higher 20°	±2 K	
5	Class 4 not fulfilled	±5 K	

3.1.3 Which Method is to Be Used for Measurement?

There are a large number of measuring methods for temperature measurement (Foken and Bange 2021a). However, fewer and fewer methods are used for reasons of measurement accuracy and handling, with platinum resistance thermometers clearly preferred for all applications a–g. Thermocouples are also used for application g, as they are particularly well suited for differential measurements. It is possible that smart sensors will be increasingly used in the future. Only for application h is temperature measurement increasingly used due to the temperature dependence of the sound velocity because of the very low time constant (→ eddy-covariance method, ▶ Sect. 3.12.3), and very thin platinum wires or thermocouples are hardly ever used. Bimetallic thermometers or thermistors are now only used in special devices.

The ***platinum resistance thermometer*** uses the temperature dependence of the electrical resistance (Harrison 2015; Foken and Bange 2021a; WMO 2024). The resistance is obtained according to the relation:

$$R(t) = R(0\,°\mathrm{C}) \cdot (1 + \alpha t + \beta t^2) \tag{3.1}$$

with the temperature t in °C and a temperature coefficient dependent on the purity of the platinum $\alpha = 0.0039083\ °\mathrm{C}^{-1}$. Under these conditions, then, $\beta = -5.775 \times 10^{-7}\ °\mathrm{C}^{-2}$. In the meteorological measurement range from –50 to 50 °C, there is a largely linear temperature dependence (▪ Table 3.5). Typically, platinum resistance thermometers are used with predominantly a nominal resistance of $R(0\ °\mathrm{C}) = 100\ \Omega$ (Pt100), sometimes also nominal resistance 1000 Ω (Pt1000).

Platinum resistance thermometers in a glass or metal tubes (▪ Fig. 3.2) in a four-wire circuit are typically used in constant current operation (▪ Fig. 3.3) with a data logger (▶ Sect. 2.7.4). It should be noted that the operating current is so low that no self-heating can take place.

Thermocouples consist of two wires of different metals, where different combinations are available depending on application and temperature range to be measured. If the two junctions of the materials have a different temperature, a current flow occurs (Seebeck effect). One junction is used as a temperature sensor, while the other is kept at a constant and known reference temperature, so that the difference between the two temperatures can be measured (▪ Fig. 3.4a). Fine wire thermocouples (▪ Fig. 3.4b) can measure temperature fluctuations in high time resolution and do not have a radiation error due to their very low mass. They are therefore suitable to calibrate or validate temperature measurements from different other methods, e.g. from sensors which are placed in unventilated radiation shields.

The temperature-dependent voltage for the most common copper-constant and thermocouple with respect to the reference temperature of 0 °C is in the relevant meteorological temperature range

$$U[\mu V] = 38.74811\ \mu V\,°\mathrm{C}^{-1} t \tag{3.2}$$

▪ Table 3.5 Maximum difference for 100 Ω platinum resistance thermometers (DIN-EN-IEC 2023)

Temperature °C	Maximum difference					
	Class A		Class B		Class AA*	
	K	Ω	K	Ω	K	Ω
−100	±0.35	±0.14	±0.8	±0.32	±0.27	±0.11
0	±0.15	±0.06	±0.3	±0.12	±0.10	±0.04
100	±0.35	±0.13	±0.8	±0.30	±0.27	±0.11

* Former class 1/3 DIN Class B

Fig. 3.2 Platinum resistance sensor. Photo © Adolf Thies GmbH & Co. KG, Göttingen, Germany

Fig. 3.3 Schematic diagram for a resistance thermometer in a four-wire circuit with constant current source (VDI 2024), with kind permission from Beuth Verlag

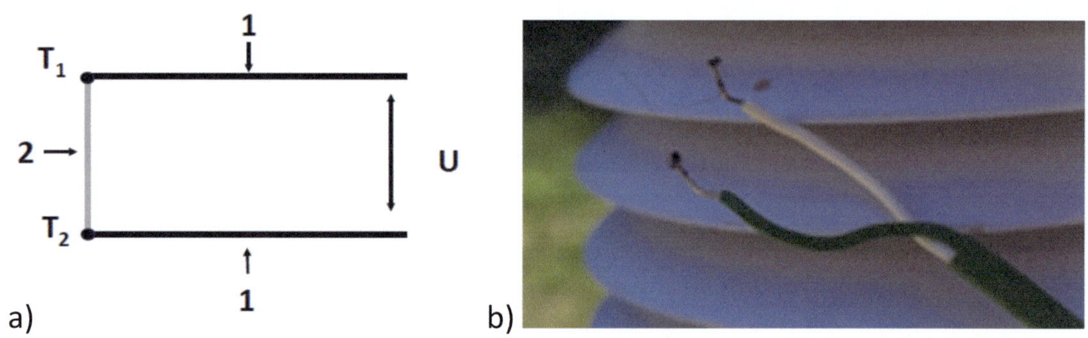

Fig. 3.4 **a** Schematic view of a thermocouple with the two temperature measurements T_1 and T_2 and the two wires of Metals 1 and 2; **b** Fine wire thermocouple (Chrome-Alumel combination) in front of a radiation shield. Photo Eitzinger, BOKU

with temperature t in °C. For other thermocouples, see DIN-EN (2009) and Foken and Bange (2021a).

Smart Sensors (environmental sensors) are the integration of electronic components on chips of only a few mm^2 in size and make it possible to integrate sensor elements together with the necessary electronics for data processing and digitization (Fig. 3.5). Among other things, the temperature dependence of semiconductor components is exploited. Sensors of this type are already used today in many technical applications (smartphones, cars, etc.). An increasing use for meteorological measurement tasks is to be expected. It is important to know that direct sensor reading can be different than

Fig. 3.5 Example of a humidity and temperature sensor. Photo Knechtel, HS Schmalkalden

the digital output to a data logger, e.g. if the raw value will be adapted by correction algorithms or factors applied (e.g. compensating a potential radiation error).

3.2 Air Humidity

Humidity measurements have been known since ancient times, using hygroscopic materials. A first instrument with good quantitative measurement was the hair hygrometer developed in 1781 by Horace Bénédict de Saussure (1740–1799), which is hardly used today for scientific applications. The psychrometer in the design according to Richard Assmann (1845–1918) from 1892 is still used as a reference instrument, the development of which goes back to John Leslie (1766–1832), dated to the year 1790. Of the multitude of measuring principles used, today almost exclusively the capacitive hygrometer and for special applications dew point hygrometers are used (Sonntag et al. 2021). Analogous to the temperature measurement, the optimal adaptation to the ambient medium air is of particular relevance.

Humidity is best expressed by the partial pressure of water vapour (water vapour pressure) e in the air, but this quantity is not measured directly. Most measuring instruments measure the relative humidity R, the ratio of water vapour pressure and water vapour pressure for saturation E at the corresponding temperature, although the water vapour pressure for saturation differs over water and ice. In addition, measuring instruments also measure the dew point τ and the absolute humidity a. In many cases, however, the derived quantities' specific humidity q and mixing ratio m are used in meteorology. The definitions and equations are given in Note 3.1.

Since the relative humidity is strongly temperature-dependent, the air temperature should always be measured in parallel to the humidity measurement. Often, sensors provide both values at the same time. Figure 3.6 shows, for example, the daily variations of relative humidity and air temperature.

Note 3.1
Humidity units (Sonntag et al. 2021)

Humidity unit	Equation
Water vapour pressure: partial pressure of the water vapour in hPa	e
Relative humidity: ratio of the water vapour pressure and the water vapour pressure at saturation in %	$R = e/E \cdot 100\%$
Dew point τ: temperature at which the water vapour pressure for saturation can be reached in °C	$E(\tau)$
Water vapour pressure for saturation with Magnus' equation (–45 to 60 °C over water) according to Sonntag (1990) in hPa	$E_w = 6.112 \cdot e^{\frac{17.62 \cdot t}{243.12+t}}$
Water vapour pressure for saturation with Magnus' equation (–65 to 0.01 °C over ice) according to Sonntag (1990) in hPa	$E_i = 6.112 \cdot e^{\frac{22.46 \cdot t}{272.62+t}}$
Absolute humidity: mass water vapour per volume moist air in kg m^{-3}	$a = \frac{0.21667 \cdot e}{T}$
Specific humidity: mass water vapour per mass moist air in kg kg^{-1} can be replaced with sufficient accuracy by the mixing ratio or vice versa	$q = 0.62198\ \frac{e}{p-0.378 \cdot e}$
Mixing ratio: mass water vapour per mass dry air in kg kg^{-1}	$m = 0.62198\ \frac{e}{p-e}$

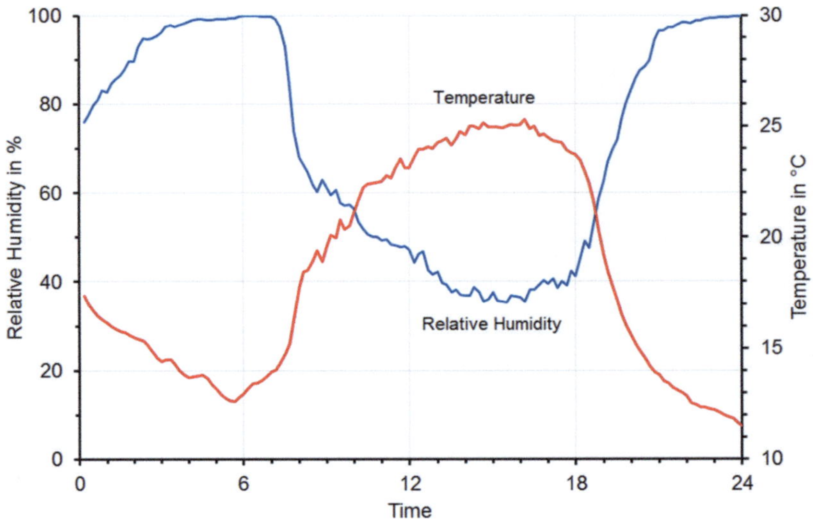

Fig. 3.6 Daily cycle of relative humidity and temperature during an arbitrary summer day (Bayreuth, Germany, Aug. 28, 2017) after Sonntag et al. (2021)

3.2.1 Why Should Air Humidity Be Measured?

Air humidity measurements are needed for various purposes in meteorology and climatology. The application determines which measuring accuracy and which time constant the sensor must have. Typical application examples are given in Table 3.6.

Table 3.7 contains the necessary accuracies and time constants for the applications. For capacitive humidity sensors, the same regulations for sensor exposure apply as for temperature measurement, see ▶ Sect. 3.1.1, especially for the temperature screen. The cap of capacitive sensors belongs to the sensor and must not be removed to reduce the time constant!

3.2.2 Where to Measure?

Since the air humidity near the surface shows considerable gradients too, the determination of a certain measuring height is necessary because of the comparability. The standard measuring height is 2 m (ISO 2015; WMO 2024). Table 3.8 contains information on the applications for which the standard measuring height is necessary or which deviating specifications apply. Information similar to Table 3.4 is not available, but it should be applied accordingly. For applications f and g, see the corresponding measurement method (▶ Sect. 3.12.3).

For measurements that are not carried out over low vegetation as in applications a–c, special specifications apply with regard to the measurement height and possible processes that may influence the measurements. Table 3.8 is applied in the analogous way, but the influencing problems listed in Table 2.1 and the atmospheric processes listed in Table 2.2 must be taken into account.

3.2.3 Which Method is to Be Used for Measurement?

For most applications (a–e), capacitive sensors are used today that measure relative humidity. For temperature-independent measurements (f) and for reference measurements, psychrometers or dew point hygrometers are used. For measurements of the latent heat flux according to the eddy-covariance method (g), only optical methods are used.

In ***capacitive humidity sensors***, the dielectric material can be either a sandwiched

Table 3.6 Application purpose for humidity measurements

Code	Application	Relevance
a	Input variable for weather, climate, and other models	
b	Climate analysis, local climate	
c	Characterization of human well-being → temperature-humidity index (heat index), Lu and Romps (2022)	
d	Determination of humidity-dependent physical quantities → e.g. air density, water vapour saturation, specific heats (Foken et al. 2021)	
e	Humidity measurements, e.g. with mobile systems for the study of the local climate → mobile measurements	
f	Determination of latent heat flux from vertical temperature gradient → see gradient measurements (There are restrictions on the choice of location and measuring equipment!)	
g	Determination of latent heat flux with the → eddy-covariance method	

Table 3.7 Accuracy of the relative humidity (capacitive sensor, g: optical sensor) and time constant for different applications (Sonntag et al. 2021) and need for a screen, if possible ventilated

Application	Accuracy	Time constant (s)	Screen	Relevance
a, b, c, d	5%	10	Yes	
e	5%	1–10	Yes	
f	5%	1–10	Yes, ventilated	
g	5% of absolute humidity	<0.01	No	

Table 3.8 Measurement height for different applications

Application	Measurement height	Relevance
a, b, c	2 m, recommended	
d	Height for which humidity measurement is required	
e	About 1–3 m	
f	Different heights between 0.25 and 10 m, logarithmical spacing	
g	Height of the sonic anemometer	

structure with two electrode surfaces on each side or it can be placed between two interleaved comb electrodes. Recently a dielectric polymer film is positioned between these (Farahani et al. 2014), where atmospheric water vapour molecules are deposited. A capacitive-type thin film humidity sensor called Humicape® (Fig. 3.7) was developed by Vaisala in Finland, and this approach is widely used in all types of humidity sensors. These sensors are usually combined with a temperature sensor and provide a voltage signal for the humidity. The smart sensor version is also common (see Fig. 3.5).

The ***psychrometer*** measures the psychrometric difference between a dry (t) and wet bulb (t_w) thermometer while both thermometers are radiation shielded and ventilated with at least 2.5 m s^{-1}, or better 3.5 m s^{-1}. The wet bulb has a wick fitting closely around the sensing element (well-washed cotton, without synthetic fibres).

The formula currently used to calculate water vapour partial pressure (Sonntag 1989) is in a simplified version

$$e = E_w(p, t_w) - \gamma(t - t_w). \tag{3.3}$$

The saturation water vapour pressure at the wet bulb thermometer (E_w) must be deter-

3

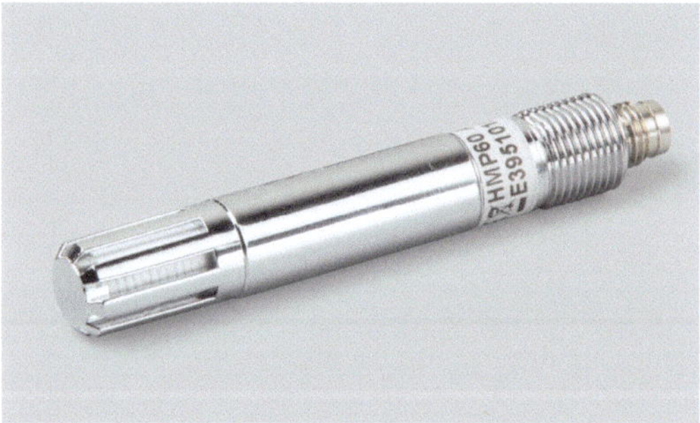

Fig. 3.7 Capacitive-type thin film humidity sensor Humicape®. Photo © Vaisala Oyj., Vantaa, Finland

mined using the formula in the Note 3.1. For iced cotton wick, the saturation vapour pressure over ice must be used. The psychrometer coefficient ($\gamma = 0.667\,\mathrm{hPa\,K^{-1}}$ for water or $0.575\,\mathrm{h\,Pa\,K^{-1}}$ for ice) is temperature- and pressure-dependent and deviates less than 5% in the usual temperature range for air pressure values above 980 hPa. For measurements in the mountains, the psychrometer coefficient must be corrected (Sonntag et al. 2021).

Figure 3.8 shows the hand-held psychrometer according to Assmann and an electrical psychrometer, as it can also be used for profile measurements (application f). The correct performance of psychrometer measurements requires some experience, since many influencing factors must be taken into account. It is essential to following the detailed instructions in the literature (Sonntag et al. 2021; WMO 2024).

Dew point and frost-point hygrometers are among the most stable and accurate instruments (Sonntag 1994), since they measure dew point or frost-point temperature directly. They are used as reference hygrometer (Sonntag et al. 2021). Most dew point and frost-point hygrometers are based on the chilled mirror principle. In these instruments, a small mirror is cooled until water condenses on its surface, either as liquid water or ice. An electronic detector senses the reflectivity of this condensate and regulates the mirror temperature such that the measured reflectivity remains constant.

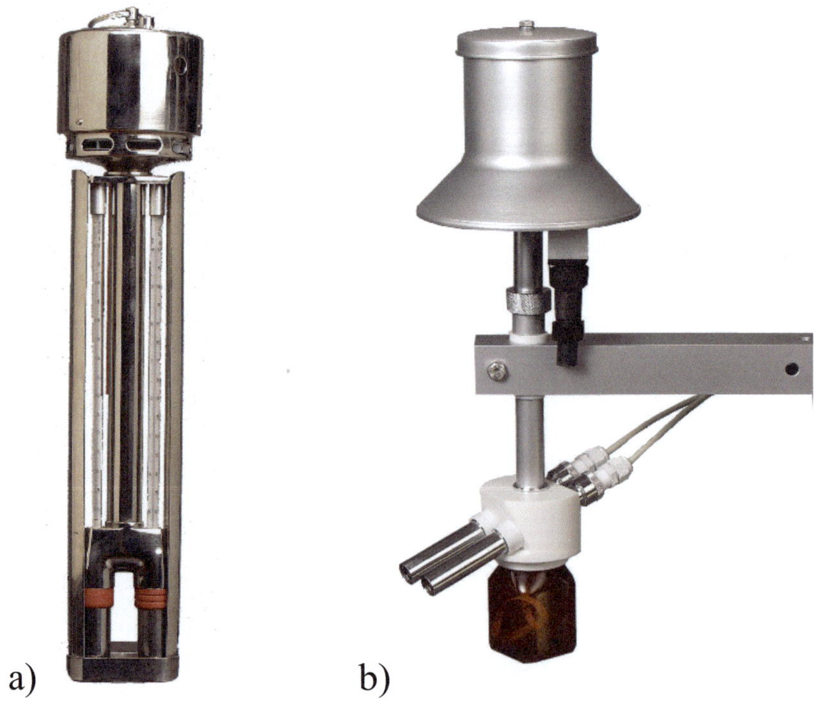

Fig. 3.8 Psychrometer according to Assmann (**a**) and Frankenberger (**b**). Photos © METEK GmbH Elmshorn, formerly Theodor Friedrichs & Co.

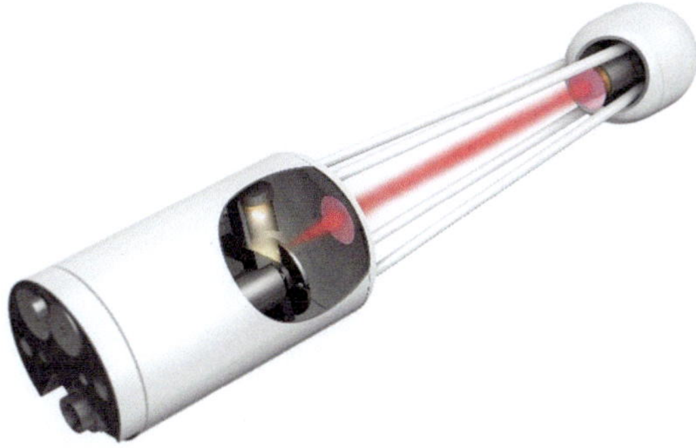

Fig. 3.9 Open-path H_2O–CO_2-gas analyser (LI-7500-DS). Photo ©, LI-COR, Inc., Lincoln NE, USA

Optical hygrometers are the only instruments suitable for measuring humidity fluctuations at high frequency up to about 100 Hz (application g). IR-hygrometers are now very sensitive and have largely replaced UV-hygrometers. They are characterized by high calibration stability. They are available as open-path instruments (Fig. 3.9) and as closed-path instruments. In the latter, the gas is aspirated at the measuring point and analysed a few meters away. In addition to water vapour, most devices also measure carbon dioxide in parallel.

The absolute humidity is calculated according to Lambert-Beer's law

$$I = I_0{}^{-akx} \tag{3.4}$$

where I is the light intensity at the receiver, I_0 the light intensity of the lamp, a the absolute humidity, k the absorption coefficient, and x the path length between lamp and receiver. Closed-path devices operate at constant temperature so that the mixing ratio of dry air and water vapour can be determined. The use of these devices, and especially their calibration, requires careful study of the operating instructions and, if necessary, instruction by an expert (Burba 2022).

3.3 Wind

The measurement of wind direction with a weathervane is already known from ancient times. However, the measurement of wind speed was a late development, since its causes were not known and were attributed to divine forces. The principle of rotational anemometer was proposed by famous Russian scientist Mikhail Vasilyevich Lomonosov (1711–1765) but realized only in 1846 by Thomas Romney Robinson (1792–1882). Theory and first measurement experiments with sonic anemometers date from the 1950s. The first sonic anemometer for measuring the three-dimensional wind vector dates from 1960 (Viktor Markovich Bovscheverov, 1905–1995), although the measurement principle is no longer used today.

Rotation anemometers (cup anemometers and propeller anemometers) and sonic anemometers are the most commonly used instruments for meteorological measurements today. Not described in the following are pitot probes for aircraft measurements, thermal anemometers partly used for military purposes, and hot-wire and laser anemometers for laboratory measurements (Foken and Bange 2021b).

3.3.1 Why Should Wind Be Measured?

Wind direction and wind speed measurements are needed for various purposes in meteorology and climatology. The application determines which measuring accuracy and which time constant the sensor must have. Typical application examples for wind speed and in most cases also wind direction are given in Table 3.9.

Table 3.10 contains the necessary accuracies and time constants for the applications. However, no time constant is specified for rotational anemometers, since this is dependent on wind speed. Instead, the distance constant is given, the product of wind speed and time constant. For measurements close to the surface (application f), low overspeeding is important, which can be realized by very sensitive instruments with a low distance constant.

3.3.2 Where to Measure?

Since the wind speed near the surface shows considerable gradients, the determination

3

Table 3.9 Application purpose for wind measurements

Code	Application	Relevance
a	Input variable for weather, climate, and other models	
b	Climate analysis, local climate	
c	Reference measurements for wind power stations (IEC 2017; Emeis and Wilbert 2021)	
d	Determination of velocity-dependent physical quantities	
e	Wind measurements, e.g. with mobile systems (during short stops) for the study of the local climate → mobile measurements	
f	Determination of turbulent fluxes from vertical temperature gradient → see gradient measurements (There are restrictions on the choice of cite location and measuring equipment!)	
g	Determination of turbulent fluxes with the → eddy-covariance method	

Table 3.10 Accuracy and distance constant of wind speed and wind direction, for different applications (Foken and Bange 2021b)

Application	Accuracy	Distance constant	Remark	Relevance
a–d	0.3–0.5 m s^{-1} 5°	2–5 m		
e	0.3 m s^{-1} 5°	2–3 m		
f	0.2 m s^{-1}	1–2 m		
g	0.1 m s^{-1}	Time constant <0.01 s	Resolution 0.01 m s^{-1}	

of a certain measuring height is necessary because of the comparability. The standard measuring height is 10 m (ISO 2015; WMO 2024). Table 3.11 contains information on the applications for which the standard measuring height is necessary or which deviating specifications apply.

The wind speed and direction measurement are also influenced by its environment. Therefore, special environmental conditions must be realized for applications a and b, or errors are possible if they are not given (Table 3.12; ISO 2015; WMO 2024). For applications c and e, the notes apply in a figurative sense and for application f and g, see the corresponding measurement method (▶ Sect. 3.12.3).

If the measuring height of 10 m for applications a and b cannot be realized due to individual obstacles, it is also possible to measure higher. There are equations on how to determine the necessary measuring height (Foken and Bange 2021b).

For measurements that are not carried out over low vegetation as in applications a, b, special specifications apply with regard to the measurement height and possible processes that may influence the measurements. Table 3.11 is applied in the analogous way, but the influencing problems listed in Table 2.1 and the atmospheric processes listed in Table 2.2 must be taken into account.

3.3.3 Which Method is to Be Used for Measurement?

For most applications (a–f), cup anemometers and possibly wind vanes are used. Today, propeller anemometers are usually only found in combination with a wind vane. Increasingly, two-dimensional sonic ane-

Table 3.11 Measurement height for different applications

Application	Measurement height	Relevance
a, b	10 m, recommended	
c	10 m and/or near the axis of the rotor	
d	Height for which wind speed measurement is required	
e	About 2–5 m	
f	Different heights between 0.25 m and 10 m, logarithmical spacing	
g	Height of the sonic anemometer	

mometers are used for all applications (Fig. 3.14c, d) and exclusively three-dimensional sonic anemometers for application g Fig. 3.14a, b).

Cup anemometers have a round or preferably conical form and consist of two parts: the rotor and the signal generator (Fig. 3.10). Cup anemometers are available with dynamo or mechanical or optical counters. The wind speed can be calculated within a certain time interval. They have a distance constant of 2–5 m. Cup anemometers have no cosine characteristic. Up to angles of about 20° between the planes of the wind field and the rotor, the correct horizontal wind is always measured (Brock and Richardson 2001).

For the calibration of cup anemometers, the knowledge of the transfer function between the speed in a wind tunnel and the speed measured with the anemometer is of high importance. This is the linear dependence between the wind speed u and the rotations n of the anemometer within a defined working range and must be determined in the wind tunnel (Kristensen 1998; Brock and Richardson 2001; DIN-ISO 2007):

$$u = a + b \cdot n. \tag{3.5}$$

The constant b is the instrument sensitivity. This calibration relation is linear over a wide range of velocities, but for low speeds (<2–4 m s^{-1}) an exponential approximation is necessary

$$u = c \cdot \exp(d \cdot n) \tag{3.6}$$

because of the threshold speed c of a rotating anemometer (about 0.1–0.3 m s^{-1}), which is

Table 3.12 Classification of wind measurement sites (ISO 2015; WMO 2024)

Class	Distance to obstacles		Angular width	Remark	Uncertainty (correction required)
	Surrounding obstacles	Thin obstacles higher than 8 m (mast, thin tree)			
1	>30 times the height of obstacles	>15 times the height of thin obstacles	≤1.9°	Obstacles lower than 4 m should be ignored	–
2	>10 times the height of obstacles	>15 times the height of thin obstacles	≤5.7°	Obstacles lower than 4 m should be ignored	≤30%
3	>5times the height of obstacles	>10 times the height of thin obstacles	≤11.3°	Obstacles lower than 5 m should be ignored	≤50%
4	>2.5 times the height of obstacles No obstacles with an angular width >60° and a height >10 m within 40 m distance		≤21.8°	Obstacles lower than 5 m should be ignored	>50%
5	Sites that do not meet the requirements of Class 4 are not recommended for wind measurements				

3

Fig. 3.10 Cup anemometer. Photo © METEK GmbH Elmshorn, formerly Th. Friedrichs & Co.

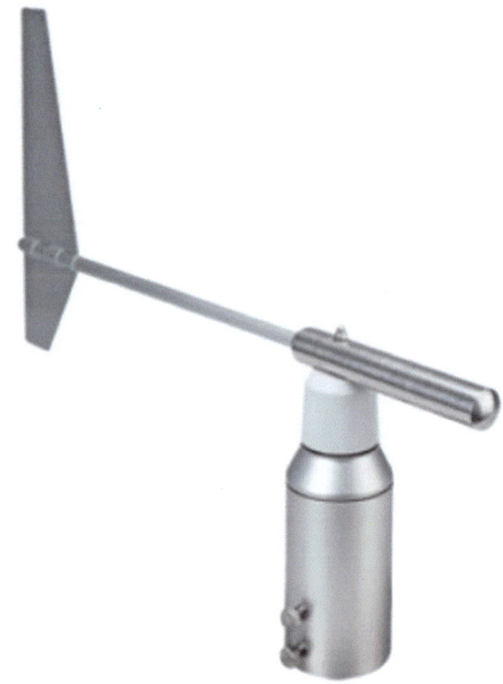

Fig. 3.12 Mechanical wind vane. Photo © Adolf Thies GmbH & Co. KG, Göttingen, Germany

the lowest wind speed that brings the rotating anemometer into continuous movement. It should not be confused with the intersection of the linearly extrapolated transfer function to the point with zero revolutions, *a*. This is illustrated in Fig. 3.11.

The ***wind vane*** is the classical instrument for measuring the wind direction (Fig. 3.12) and is a dynamical system of the second order. This means that they adjust to the correct wind direction under oscillations. The damping of the oscillations is an important measure (DeFelice 1998; Brock and Richardson 2001). The wind vane consists of a wind direction indicator rotatable about a vertical axis. The wind direction is given in degrees (N = 0°, 360°, E = 90°, etc.). A direct averaging of the wind direction is not possible because of the jump at north. Corresponding procedures are given in the literature (Foken and Bange 2021b).

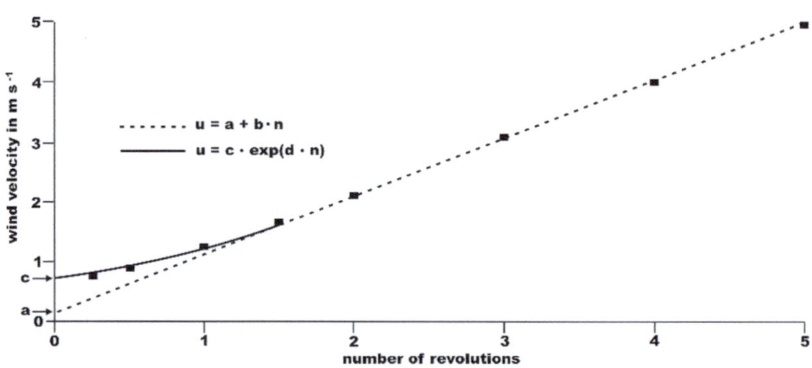

Fig. 3.11 Transfer function of a cup anemometer with*n*: number of revolutions of the anemometer; *u*: wind velocity; and *a*, *b*, *c*, and *d*: constants, after (VDI 2018) with permission from VDI e.V, Düsseldorf, Germany

Propellers have a helicoidal form that ensures a near cosine response that is the ratio of the measured wind speed for a special angle α of incidence to the wind speed of the horizontal wind field $u(0)$ multiplied by the cosine of the angle

$$F(\alpha) = \frac{u(\alpha)}{u(0) \cdot \cos\alpha} \tag{3.7}$$

An ideal cosine response is given by $F(\alpha)=1$. In the past, this was even exploited by measuring the vector of turbulent wind fluctuations with three propellers. Today, propellers are mainly used in conjunction with a wind vane (◻ Fig. 3.13). The cosine dependence does not play a role, since the propeller is always ideally orientated.

Initially, ***sonic anemometers*** had only a three-dimensional design, and the measurement paths of the sonic anemometers were predominantly Cartesian-orientated. They were mainly applied for flux measurements (◻ Fig. 3.14a, b; see ▶ Sect. 3.12.3). Modern sonic anemometers have increased angles of the measurement paths to obtain lower flow distortion. Technical progress in the last 10–20 years has reduced the costs dramatically, and the measuring principle is also available for two-dimensional sensors that measure wind speed and direction for general purpose (◻ Fig. 3.14c).

Modern sonic anemometers use the travel time principle and a direct time determination (Hanafusa et al. 1982). In this method, a sonic signal (about 100 kHz) is transmitted from both sides of a measurement path and received on the opposite sides. Due to the wind velocity, one signal is faster than the other. The exact travel times of the sonic signals

$$t_{1,2} = \frac{\sqrt{c^2 - u_n^2} \pm u_d}{c^2 - u^2} d \tag{3.8}$$

are used for the determination of the wind velocity, where d is the path length, u_d is the wind component along the path, u_n is the normal component of the wind, and c is the sound speed. The calculation of wind speed and sonic temperature is described in Note 3.2.

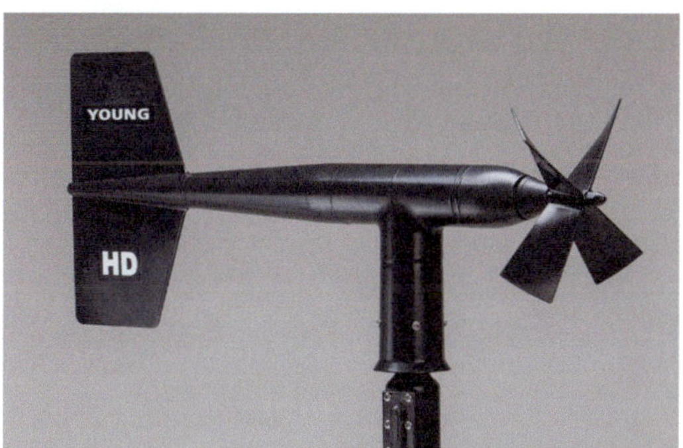

◻ **Fig. 3.13** Combination of propeller anemometer and mechanical wind vane, also called skyvane. Photo © R. M. YOUNG Company/GWU-Umwelttechnik GmbH, Erftstadt, Germany

Note 3.2

Calculation of wind speed and sonic temperature

The difference of the reciprocal travel times gives the wind velocity along the measurement path

$$\frac{1}{t_1} - \frac{1}{t_2} = \frac{2}{d} u_d \tag{N3.1}$$

In practice, the detection limit is often given by the zero point drift of the speed measurement that can be verified in a null wind chamber (DIN-ISO 2003). The transducers and the mounting rods cause deformation of the wind field. These effects can be determined through wind tunnel calibrations. However, this shadow error is usually lower in the turbulent wind field (see also ▶ Sect. 3.10).

The sum of the reciprocal travel times gives the sound speed

$$\frac{1}{t_1} + \frac{1}{t_2} = \frac{2}{d} c \sqrt{1 - \frac{u_n^2}{c^2}} \approx \frac{2}{d} c \tag{N3.2}$$

which is a function of the temperature and the moisture (e: water vapor pressure; p: pressure; q: specific humidity) and can be converted to the sound temperature (Foken and Bange 2021a)

$$T_s = T(1 + 0.531q) \approx T\left(1 + 0.33\frac{e}{p}\right), \tag{N3.3}$$

3

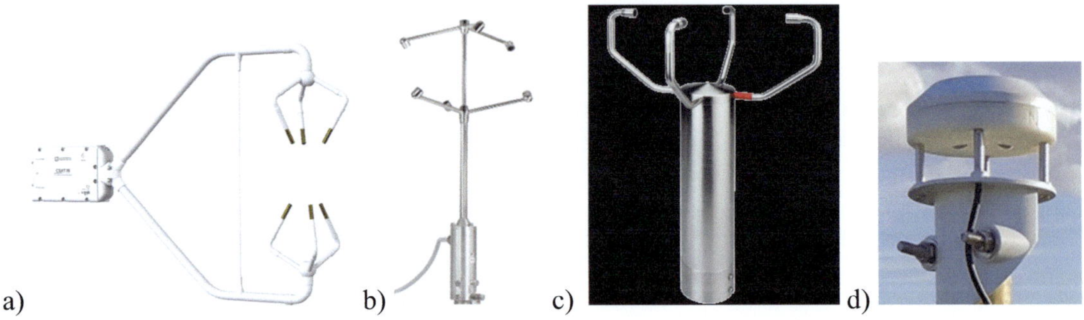

Fig. 3.14 Sonic anemometers: **a** orientated three-dimensional sonic anemometer CSAT3B. (Photo © Campbell Sci. Inc. Logan, UT, USA); **b** omnidirectional three-dimensional sonic anemometer uSonic-3 Scientific (former: USA-1; Photo © METEK GmbH, Elmshorn, Germany); **c** Two-dimensional sonic anemometers (Photo © Adolf Thies GmbH & Co. KG, Göttingen, Germany); **d** Two-dimensional DS-2 Sensor (DECAGON DEVICES) (Photo: Eitzinger, BOKU)

which is similar to the virtual temperature (see Note 3.3).

Note 3.3

Virtual temperature

The virtual temperature T_v is the temperature of the dry air, which has the same density as the moist air

$$T_v = T(1 + 0.61q) \approx T\left(1 + 0.38\frac{e}{p}\right), \qquad \text{(N3.4)}$$

The virtual temperature is always used when the density changes noticeably due to the water vapour content in the air, and the process to be investigated depends on the air density. For symbols see Note 3.1.

3.4 Atmospheric Pressure

Air pressure measurements are essential for weather forecasting; for many micrometeorological applications, a less precise measurement is usually sufficient, e.g. to determine the air density. Pressure measurements are one of the oldest in meteorology. The first liquid barometer was built in 1643 by Evangelista Torricelli (1608–1647). Today, these are no longer allowed to be used due to the toxicity of mercury. Aneroid barometers, first constructed in 1844 by Lucien Vidie (1805–1866), are still widely used today—but no longer in the scientific field. Today, electronic sensors are used, often integrated on a chip (environmental sensor).

3.4.1 Why Should Atmospheric Pressure Be Measured?

The air pressure is not measured directly in many micrometeorological experiments if a pressure sensor is not integrated in another measuring device. It is often sufficient to determine the pressure using the barometric equation (Note 3.4). The normal pressure (1013 hPa at sea level) or, with greater accuracy, the air pressure at a nearby meteorological station can be used. Typical application examples are given in Table 3.13. Table 3.14 contains the necessary accuracies for the applications.

Note 3.4

Barometric equation

The pressure at the height z in m can be calculated from the pressure at sea level $p(z=0)$ and the mean virtual temperature, T_v in K (see Note 3.3), between sea level and z, with the constant gravity acceleration $g_0=9.81$ m s^{-2}, and the gas constant of dry air $R_L=287.0586$ J kg^{-1} K^{-1}

Table 3.13 Application purpose for humidity measurements

Code	Application	Relevance
a	Input variable for weather, climate, and other models	
b	Determination of pressure-dependent physical quantities → e.g. air density (Foken et al. 2021)	

Table 3.14 Accuracy of pressure measurements for different applications

Application	Accuracy (hPa)	Remark	Relevance
a	0.1–0.2		
b	About 5	Note 3.4: With constant pressure at sea level, the difference to the actual pressure should be added	

$$p(z) = p(z = 0)\, e^{\frac{g_0}{R_L \cdot T_v}} \tag{3.5}$$

In the lower troposphere, the geopotential height that is actually required can be replaced by the geometric altitude, which differs only slightly, as is the case here. In addition, depending on the required accuracy, the virtual temperature can be replaced by the actual temperature. The mean temperature between the measurement altitude and the sea-level altitude can be determined by taking into account the temperature decrease of 0.6 K per 100 m.

3.4.2 Where to Measure?

For application a, the pressure must be measured at the height of the measuring station specified in the station documentation. For application b, the measuring height is not critical due to the low accuracy required.

3.4.3 Which Method is to Be Used for Measurement?

Today, electronic sensors are used almost exclusively (Torri et al. 2021). The silicon capacitive pressure sensors with high sensitivity, good long-term stability, and low temperature dependence are mainly used in application a when an accuracy of 0.1 hPa is required. In principle, however, the slightly less accurate and cheaper silicon piezoresistive pressure sensors can also be used. They are very small (environmental sensors) and have good linearity and are suitable for applications a and b. For application b, it is often sufficient to determine the air pressure using the barometric equation (Note 3.4). The user should check the accuracy required for the air density, for example, and can find the necessary pressure dependence in Foken et al. (2021).

3.5 Shortwave and Longwave Radiation

Radiation measurements have been possible for around 200 years. The first pyrheliometer for measuring direct solar radiation with an aperture angle of 5° was constructed by Claude Servais Mathias Pouillet (1790–1868) in 1838. Hugh Longbourne Callendar (1863–1930) developed a device for measuring shortwave solar radiation from the upper half-space (pyranometer), which was apparently used for the first time in 1906. The first pyrgeometer for measuring longwave terrestrial radiation was developed by Knut Ångström (1857–1910) in 1905. In the International Geophysical Year 1957/58, radiation measurements were intensified and around 1990, with the Baseline Surface Radiation Network (BSRN) project, considerable progress was made in sensor technology, and accuracy and radiation measurements are widely used today. Behrens (2021) provides a comprehensive overview of radiation measurement, including all the basic principles and measuring devices.

The categorization of radiation is shown in Note 3.5. Shortwave solar radiation is assigned to the range from 0.28 µm to 3.0 µm and longwave terrestrial thermal radiation to the range from 3.0 to 100 µm. The measuring instruments, which are all relative instruments, are categorized accordingly, whereby the incident radiation is compared with that on a reference surface. Absolute instruments are only used in calibration institutes. In addition to the wavelength, a distinction is made between instruments that measure the radiation from a half-space or are only directed at the sun or for measuring the surface temperature.

Note 3.5

Spectral classification of the radiation

Spectral classification of short- and long wave radiation (ISO 2007; Wendisch and Yang 2012, modified)

Notation	Wavelength in µm	Remarks
Ultraviolet radiation		
UV-C-range	0.010–0.280	Does not penetrate the atmosphere
UV-B-range	0.280–0.315	Does partly penetrate the atmosphere
UV-A-range	0.315–0.380	Penetrates the atmosphere
Visible radiation		
Violet	0.380–0.455	
Blue	0.455–0.492	
Green	0.492–0.576	
Yellow	0.576–0.597	
Orange	0.597–0.622	
Red	0.622–0.680	
Dark red	0.680–0.780	
Infrared radiation		
Near infrared (NIR), IR-A	0.780–1.4	
Near infrared (NIR), IR-B	1.4–3.0	
Mean infrared (MIR), IR-C	3.0–50	
Far infrared (FIR), IR-C	50–1000	

▫ Table 3.15 provides an overview. Pyranometers (0.28–3.0 µm) and pyrgeometers (5.0–100 µm, the range 3.0–5.0 µm has only little energy and is taken into account in the calibration; see ▸ Sect. 1.1.1) or devices that measure both spectral ranges and the upper and lower half-space together (net radiometers) are mainly used for the following applications. There are also PAR sensors that measure in the photosynthetically active range from 0.4 µm to 0.7 µm as well as measuring devices that only measure certain spectral ranges, e.g. in the UV. Sensors that measure direct and diffuse solar radiation and calculate the global radiation from this are also widely used.

3.5.1 Why Should Radiation Be Measured?

Radiation fluxes are needed for various purposes in meteorology and climatology. The application determines which measuring accuracy and which time constant the sensor must have. Typical application examples mainly for global radiation and net radiation are given in ▫ Table 3.16.

▫ Table 3.17 contains the necessary accuracies and time constants for the applications. There is a classification of sensors for shortwave radiation (pyranometers).

3.5.2 Where to Measure Radiation?

There are no specifications for the measurement height for radiation measurements, es-

Table 3.15 Classification of radiation sensors (Foken and Mauder 2024)

Measurement device	Wavelength		Opening angle	
	Short wave	Long wave	Half-space	Directed
Pyrheliometer	x			x
Pyranometer	x		x	
Albedometer	x		x*	
Pyrgeometer		x	x	
Net pyrgeometer		x	x*	
Radiometer	x	x	x	
Net radiometer	x	x	x*	
PAR radiation sensor	x (partly)		x	
IR radiation thermometer		x		x

* Upper and lower half-space

Table 3.16 Application purpose for radiation measurements

Code	Application	Relevance
a	Input variable for weather, climate, and other models	
b	Climate analysis, local climate	
c	Reference measurements for photovoltaic stations (Emeis and Wilbert 2021)	
d	Radiation measurements, e.g. with mobile systems for the study of the local climate→mobile measurements (e.g. (during short stops)	

Table 3.17 Accuracy and time constant of pyranometers (ISO 2018)

Application	Accuracy ($W\ m^{-2}$)	Time constant (s)	Pyranometer class	Relevance
a, b	±10	<10	A	
a, b, d	±20	<20	B	
c, d	±30	<30	C	

pecially for short- and longwave radiation from above. Here, the WMO only specifies requirements for the prevention of possible shadowing, and reflective objects with an albedo value greater than 0.5 should be avoided (ISO 2015; WMO 2024). Table 3.18 provides an overview of the data quality classes.

For the radiation fluxes from below, as required for measuring the radiation balance, an estimate according to Latimer (1972) applies. For a device installed at a height of 2 m, 90% of the radiation comes from a circle with a diameter of 12 m and 95% (99%) with a diameter of 18 m (40 m). At a height of 4 m, the diameters double. When setting up the net radiometers, it is therefore important to ensure that there is a uniform surface (temperature, surface humidity, vegetation) in the area of influence of the net radiometer. If this is not the case, the error can be determined from deviations in the area of influence. Similar criteria should be applied for longwave radiant fluxes.

3

Table 3.18 Classification of global and diffuse radiation measurement sites (ISO 2015; WMO 2024), ϕ: geographic latitude; h_S: altitude of sun

Class	Shading obstacles		Reflecting obstacles
	ϕ < 60°	ϕ ≥ 60°	
1	$h_S > 5°$	$h_S > 3°$	No nonshading reflecting obstacles with angular height above 5° and total angular width above 10°
2	$h_S > 7°$	$h_S > 5°$	No nonshading reflecting obstacles with angular height above 7° and total angular width above 20°
3	$h_S > 10°$	$h_S > 7°$	No nonshading reflecting obstacles with angular height above 15° and total angular width above 45°
4			No shade for more than 30% of daytime, or any day of the year

3.5.3 Which Method is to Be Used for Measurement?

The measuring devices to be used differ in terms of the spectral range to be measured. For many micrometeorological measurements, the use of a net radiometer is recommended, which contains pyranometers for the shortwave radiation from above and below and pyrgeometers for the longwave radiation from above and below.

Pyranometers are used for measurements in the upper half-space to record shortwave downwelling radiation and in combination with a shadow ring for shortwave diffuse radiation. Ventilation of the dome can prevent dew formation in the morning. For measurements from the lower half-space, the shortwave upwelling (reflected) radiation is measured. The combination of one device pointing upwards and one pointing downwards is an albedometer. Pyranometers are also used in an inclined position, e.g. for monitoring photovoltaic systems.

Measuring elements are often thermocouples which are located in white and black areas. However, photodiodes are also used. The measuring elements are protected against environmental influences by a glass or quartz dome, which limits the spectral transmittance.

Pyranometers are often calibrated by the manufacturer or national laboratories at the meteorological services. The devices are categorized into classes A, B, and C with regard to many specifications (ISO 2018), whereby Class B is sufficient for most applications. If values deviating from zero are nevertheless displayed during the night, they must be set to zero.

Pyrgeometers are used for measurements in the upper half-space to measure the longwave downwelling radiation. For measurements from the lower half-space, the longwave upwelling thermal radiation is measured. Since pyrgeometers measure the difference between the longwave radiation of the device and the longwave radiation from above and below, they have a temperature measurement of the inner housing. To avoid overheating, pyrgeometers should be ventilated. The measured signal is simply the difference between the incoming longwave radiation and the longwave radiation of the housing in the form

$$I \downarrow = \frac{U_{\mathrm{rec}}}{C} + k\sigma_{\mathrm{SB}} T_G^4, \qquad (3.9)$$

where U_{rec} is the voltage measured at the receiver, C and k are calibration coefficients, T_G is the housing temperature which must be measured, and $\sigma_{SB} = 5.67 \times 10^{-8}$ W m^{-2} K^{-4} is the Stefan-Boltzmann constant. With very good pyrgeometers, the temperature in the silicon dome is also measured. This leads to more complex evaluation equations (Behrens 2021; Foken and Mauder 2024). The measuring elements are similar to pyranometers. There is no quality classification of the devices. However, good devices measure with comparable accuracies.

Checking the longwave radiation fluxes is not easy because errors in the housing temperature must also be taken into account.

Simple test options are given below (Gilgen et al. 1994). For the upwelling longwave radiation, the test is to see if the radiation is within a certain difference of the radiation according to Stefan-Boltzmann law with the housing temperature of the sensor:

$$\sigma(T_G - 5K)^4 \leq I \uparrow \leq \sigma_{SB}(T_G + 5K)^4. \tag{3.10}$$

For strong longwave upwelling radiation during the night or strong heating of a dry soil, the threshold value of 5 K must be increased. For the atmospheric downwelling longwave radiation, the test is related to a black body (dome covered with water) and a grey body (clear sky with a radiation temperature of –55 °C):

$$0.7\ \sigma_{SB}\ T_G^4 \leq I \uparrow \leq \sigma_{SB}\ T_G^4. \tag{3.11}$$

Radiation thermometers are pyrgeometers that have only a small aperture angle and measure in atmospheric windows of 8–12 µm. The surface temperature is determined using the Stefan-Boltzmann law. They are also available in the form of thermal imaging cameras.

Net radiometers measure the short- and longwave radiation from the lower and upper half-space, using two pyranometers and two pyrgeometers as well as a thermometer for the housing temperature. It is strongly recommended to use only devices with four separate sensors and ventilation. Simple devices (radiometers) that measure shortwave and longwave radiation together are not recommended due to lower accuracy, e.g. the radiation balance can be underestimated (Brotzge and Duchon 2000). A corresponding device is shown in ◘ Fig. 3.15.

PAR radiation sensors measure the radiation in the photosynthetically active frequency range from 400 to 700 nm. The radiation is measured in µmol $m^{-2}\ s^{-1}$ photon flux and corresponds to approximately half the value of the downwelling shortwave radiation in W m^{-2}, which can be used for control purposes, as the devices can become dirty relatively quickly and have to be recalibrated.

3.6 Precipitation

Precipitation measurements have been known since ancient times. The first work on this was done by Beneddetto Castelli (1577–1643) in Italy. Improved instruments were built by Giovanni Poleni (1683–1761) in Padova and Paride M. Salvago in Genova. The first tipping-bucket rain gauge was built by Sir Christopher Wren (1632–1723) in England in 1679. In the centuries that followed, there was a multitude of device developments. At the end of the last century, the WMO endeavoured to establish guidelines for precipitation measurements through comparison experiments (Cauteruccio et al. 2021).

Precipitation occurs in various liquid and solid forms, which are explained in Note 3.6. Precipitation can be long-lasting or short-lived (showers).

Note 3.6
Various liquid and solid forms of precipitation

Form of precipitation	Description
Rain	Water droplets with a diameter of about 0.5–5 mm
Drizzle	Water droplets with a diameter of about 0.1–0.5 mm
Snow	Solid precipitation of individual or adhering ice crystals
Snow drizzle	Sleeted cloud droplets or snow stars less than 1 mm diameter
Soft hail	Mostly rounded grains of snow-like texture with a diameter of 2–5 mm
Hail	Falling ice particles in general with diameters between 5 and 50 mm, in extreme cases over 100 mm

Measuring devices are used to determine whether a precipitation event is taking place without specifying the amount (rain detector), to determine the amount of rain or

3

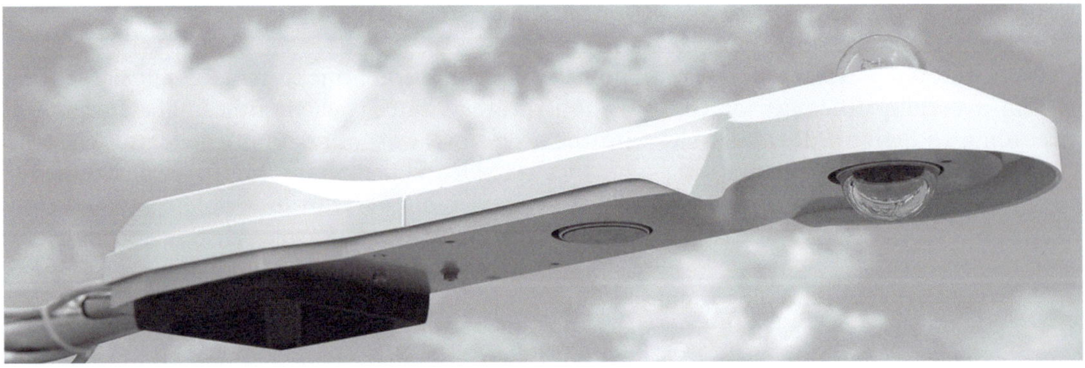

Fig. 3.15 Net radiometer CNR4 with two as albedometer installed pyranometer (right) and pyrgeometer (middle) and a common ventilation. Published with kind permission of © OTT HydroMet, All Rights Reserved

snow (the total volume of liquid or solid precipitation deposited in a given time interval per unit area of the horizontal projection of the ground surface) measured in mm and the intensity of liquid or solid precipitation measured in mm h^{-1}.

All precipitation gauges have two typical errors: The wetting error due to evaporation losses from wetted surfaces and the wind error due to losses during measurement. In the case of liquid precipitation, up to 10% less precipitation is determined; in the case of solid precipitation, it can be significantly more. There are suggestions for correction, but these are not routinely applied. This means that all reported precipitation measurements are subject to errors, which must be taken into account when calculating the water balance.

3.6.1 Why Should Precipitation Be Measured?

Precipitation is a key variable in the water balance equation and must be measured in all hydrological analyses. However, it is also closely related to evaporation, so it is also important in energy analyses. However, no experiment should be without precipitation measurements in order to be able to interpret data correctly. There are the following main areas of application (Table 3.19).

Table 3.19 Application purpose for precipitation measurements

Code	Application	Relevance
a	Input variable for weather, climate, and other models	
b	Reference data for experiments, etc.	
c	Hydrological measurements	

3.6.2 Where to Measure Precipitation?

Measuring devices for precipitation generally record this at 1 m above ground. If greater snow depths are expected, measurements can also be taken at 1.5 to 2.0 m above ground. The WMO has made special stipulations to exclude the influence of obstacles but also to reduce wind speeds (ISO 2015; WMO 2024). Table 3.20 provides an overview of the data quality classes.

3.6.3 Which Method is to Be Used for Measurement?

There are many direct and indirect measurement methods for precipitation, whereby this book is limited to the most common direct measurement methods; for further details see Cauteruccio et al.(2021). Optical or acoustic

Table 3.20 Most important criteria for the quality assessment of precipitation measurements (ISO 2015; WMO 2024)

Class	Flat terrain, slope	Obstacles, elevation angle	Additional uncertainty	Application
1	<19°	26,5°, at a distance two times their height 14°, at a distance four times their height		a–c
2	<19°	26,5°, at a distance two times their height	+5%	b
3	<30°	45°, at a distance of their height	+15%	(b)
4	>30°	45°, at a distance of their height	+25%	
5		Closer one half of their height	+100%	

a)

b)

Fig. 3.16 Schematic representation of the **a** tipping-bucket gauge and **b** weighing measuring principle. From Cauteruccio et al. (2021) with kind permission of Springer

distrometers are also mainly used in research to determine precipitation intensity and droplet size (Emeis 2010). Area-wide precipitation measurements are carried out using weather radar (Seltmann 2021).

The simplest ***catching-type gauge*** appliances only have a standardized collection surface and a collection container. These are usually emptied daily, in most countries around 7 am. They are suitable if neither an exact time allocation of the precipitation event nor a determination of the intensity is necessary. It is the easiest way to determine precipitation in simple experiments and climatic studies.

The ***tipping-bucket gauge*** is a very common method of precipitation measurement and is also used in simple weather stations for private use, so that these can also be used for precipitation monitoring (application b). The measuring principle is shown in Fig. 3.16a. The precipitation water falls into a bucket, which tips after filling and empties again. The number of emptying cycles is counted electronically. Thanks to many years of experience, these devices are very accurate, but are not suitable for high precipitation intensities and must be heated in frosty conditions.

The ***weighing gauge*** has no moving measuring elements and is suitable for liquid and solid precipitation and also for high intensities. The measuring principle (◘ Fig. 3.16b) consists of a collecting vessel which is placed on a scale, whereby the weight is measured electronically. The devices must be emptied automatically or during periods without precipitation. Possible evaporation can be prevented by an oil film. Devices often have incorrect measurements in very strong winds.

The ***drop counter gauge*** is similar to the tipping-bucket gauge, but the precipitation water is channelled through a calibrated drop counter and the drops are counted electronically. The device must also be heated in frosty conditions and is not suitable for high intensities. It requires a high level of maintenance.

3.7 Leaf Wetness and Dew

Leaf wetness refers to the presence of liquid water on the leaf surface. There are four sources of water for the surface of leaves: (i) precipitation (interception), (ii) overhead irrigation, (iii) dew, and (iv) guttation (driven by leaf water potential). Three quantities commonly used to describe leaf wetness include the amount of water retained per unit leaf area, the portion of the leaf covered by water, and the duration of leaf wetness. The properties of leaf wetness primarily depend on plant-specific characteristics, including leaf area and angle, surface wettability, and meteorological conditions. They can significantly affect the water balance of individual plants and canopy. The maximum amount of water that can be retained per unit leaf area before it drips off varies among species, ranging from 0.1 to 500 ml m^{-2} (Lalic et al. 2018).

Dew results from the direct condensation of atmospheric water vapour on underlying surfaces and objects at temperatures above 0 °C. A collected dew from dense canopies of field crops or grass rarely exceeds 0.5 mm, though, theoretically, it can measure up to 0.8 mm. At low temperatures, dew freezes, producing window frost.

3.7.1 Why Should Leaf Wetness and Dew Be Measured?

The presence or absence of liquid water on the leaf surface influences plant disease development, transpiration intensity, and overall plant-atmosphere gas exchange. The duration and amount of leaf wetness play significant roles in the energy and water balance of the plant canopy and the air surrounding it. In small amounts plant leaves can take up water through the leaf surface (especially stomata) from wet surface or release water under high leaf water potential (guttation). Therefore, accurately determining these factors is of utmost importance from both the theoretical and practical perspectives of plant physiology and pathology and understanding the microscale conditions of the surface atmospheric layer. ◘ Table 3.21 gives typical application examples.

3.7.2 Where to Measure Leaf Wetness and Dew?

To ensure measurements that accurately represent leaf wetness conditions across a large area, leaf wetness sensors must be shielded from radiation (especially solar) and wind exposure due to the significant impact on the deposition and evaporation of liquid water on the leaf surface. Additionally, it should be ensured that the temperature of the leaf wetness sensor does not differ from its environment (i.e. the leaves). The sensor should not be placed directly on another thermally stable surface, e.g. onto the measuring pole or an arm, and it should be kept clean. It is also important to consider the existence of different leaf wetness gradients resulting from precipitation interception and water vapour condensation. In a study of a soybean canopy, Schmitz and Grant (2009) measured using re-

Table 3.21 Application purpose for leaf wetness and dew measurements

Code	Application	Relevance
a	Plant diseases monitoring and forecasting	
b	Input variable for soil-vegetation-atmosphere-transfer (SVAT) models	
c	Operational parameter in deposition models (trace gases and aerosol particles) at least on semiquantitative basis	
d	Agricultural irrigation and growth management	

sistance sensors that the canopy top experienced longer periods of wetness from condensation compared to precipitation interception. In the middle of the canopy, the frequency of wetting from condensation exceeded that from rain, but wetness due to rain lasted twice than wetness due to condensation. At the canopy's bottom, wetness episodes due to condensation were rarely recorded.

In general, leaf wetness sensors should be installed above soil surface, fixed on stems or leaves (clip sensors) within the canopy or tree crown facing upwards, with slope and orientation that mimics leaf surface slope. The specific siting of a sensor within a specific canopy should meet the problem investigated (e.g. the area of deseases occurring).

3.7.3 Which Method is to Be Used for Measurement?

If *historical data* about leaf wetness are available, it is important to note that leaf wetness measurement methods changed over time. First, mechanical leaf wetness sensors which were in use until the 1970s recorded changes in sensing material size, length, or weight due to the adsorption of water retained on the surface. The following generation of leaf wetness sensors was deployed, using measurements by filter paper conductivity variation as a function of wetness. Filter paper as a sensing surface has several shortcomings, of which the most important are that its wetting and drying mechanisms differ from natural leaves and that exposure to solar radiation, high wind speeds, or high temperatures changes its physical characteristics related to wetting and drying. Other methods use different sensing surfaces which are supposed to mimic leaf surface (Fig. 3.17). However, the concept "one-size-fits-all" completely omits differences in shape, texture, surface structure (such as hairs) and hydrophobicity among plant leaves which is still an issue while measuring leaf wetness.

Classification of leaf wetness sensors is typically based on a method for detecting and measuring the presence of water on the leaf surface. All leaf wetness sensors indirectly measure selected (non-leaf wetness) quantities by mimicking the leaf surface and measuring changes in the electrical conductivity (resistance) or capacity of the sensor's surface when water is present.

Recently designed paper-based chipless sensors detect shifts in the water's dielectric permittivity at specific microwave frequency and temperature (Dey et al. 2020). Commercially available sensors almost without exception work on the capacitive principle (Rowlandson et al. 2014). Two such types of instruments are shown in Fig. 3.17 representing standalone devices or elements of automated weather station.

The listed measurement techniques introduce two primary sources of uncertainty in leaf wetness measurements: (i) the ability of leaf wetness sensors sensing surface to accurately replicate the physical and chemical characteristics of a leaf's surface, and (ii) the representativeness of the physical model along with its technological implementation within the sensor design.

Significant step forward is made with recent design of the Bio-mimetic leaf wetness sensor from replica moulding of leaves (Nguyen et al. 2023). This sensor is able to measure leaf wetness duration as actually ex-

3

a)

b)

Fig. 3.17 Two different design examples of leaf wetness sensors: **a** type Campbell Scientific and **b** type Phytos 31, METER. Photos Eitzinger, BOKU

perienced on leaves of diverse plant species, because of its surface, that is the exact replica of a leaf with the same wetting properties. The sensor consists of a circuit board with a printed inter-digitated capacitive element made of copper, which is covered by two layers of polydimethylsiloxane (PDMS).

3.8 Soil Temperature and Soil Heat Flux

Initially, measurements and modelling in the soil mainly concerned water flows in connection with the studies by Henry Darcy (1803–1858) and in the last century by Lorenzo A. Richards (1904–1993); see also ▶ Sect. 3.9. The laws of physics also apply in principle to heat fluxes. Measurements of near soil surface temperatures have been part of the measurement program since the beginning of regular meteorological measurements in the second half of the nineteenth century (Huwe et al. 2021). However, these are measured in a monolith of sandy soil and are only comparable with natural soil to a limited extent. The soil temperature as a part of the heat balance of the Earth's surface was already discussed in textbooks of this time. With the beginning of comprehensive surface energy studies in the 1950s–1970s and more accurate radiation measurements, the interest to measure the ground heat flux increased (Sauer and Horton 2005).

3.8.1 Why Should Soil Temperature and Soil Heat Flux Be Measured?

The soil is the lower boundary condition of the atmosphere and thus determines the stability of the atmosphere. The soil is also an energy sink or source, and therefore the soil heat flux is part of the energy balance at the earth's surface (see ▶ Sect. 1.1.4). The soil properties (temperature and moisture) are decisive for the evaporation process. Monitoring soil properties, at least soil temperature, is therefore essential for many of the applications shown in ◘ Table 3.22.

Table 3.22 Applications for soil temperature and soil heat flux measurements

Application	Temperature	Heat flux	Description	Relevance
a	x		Input variable for weather, climate, and other models and studies	
b	x		Reference data for experiments, etc.	
c	x	x	Energy balance studies	
d	x	(x)	Control measurements for systems for heat recovery from the ground	

3.8.2 Where to Measure Soil Temperature and Soil Heat Flux?

The soil temperature measurements are mainly taken in the upper layer, which is around 1 m thick, as the changes below this are only slight and occur very slowly (see ▶ Sect. 1.1.3). The aim is to achieve an approximately logarithmic distribution of sensors with depth, with the sensor density being highest near the surface. Typical measurement depths are 5, 10, 20, 50, and 100 cm. Sometimes measurements are also taken at a depth of 2 cm, but this depth can often not be exactly maintained, e.g. under a meadow (Bonan 2019; Huwe et al. 2021).

The soil heat flux should be measured as deeply as possible, i.e. at a depth of about 20 cm, so that unavoidable errors have less influence on the heat flux to be determined at the soil surface due to the lower heat flux. It can be measured both with soil heat flux plates and from the temperature gradient between 15 and 25 cm depth (two additional temperature measurements). To record the heat storage in the soil layer above the heat flux measurement, a compression of the temperature measurements is recommended. In any case, soil moisture measurements are necessary to determine the heat capacity of the soil (Liebethal et al. 2005); see also ▶ Sect. 3.8.3.

The installation of the sensors in the soil must fulfil several competing requirements (see also ▶ Sect. 3.9):

- The area to be measured should be representative of the (undisturbed) soil, i.e. avoid worm and root passages as well as unusually large or numerous stones that restrict the water flow or focus the heat flux.
- The spatial position of the individual sensors and their positions in relation to each other should be known down to a few centimetres.
- The sensors should be installed in such a way that they are never disturbed, have direct contact with the natural ground, and are backfilled in such a way that the filling, preferably with existing material, has only a negligible effect on the water and heat flows.
- It must be ensured that the soil surface above and around the sensors remains permanently permeable throughout the measurement period (representing the undisturbed conditions to be measured).
- Any tension, pressure, or vibration on the supply lines to the sensors must be and remain excluded.

Installation can be carried out from the soil surface or through the side wall of a profile pit (Op de Beeck et al. 2017; Wessel-Bothe and Weihermüller 2020). When installing individual sensors from a profile pit, it can be difficult to fill the borehole for the probes if the borehole was not created with a sufficient slope. In the case of installations very close to the profile pit wall, unrepresentative loosening of the soil that cannot be levelled out again cannot be ruled out.

In recent years, more and more compact solutions have been offered that are easy to install and contain water content and temperature sensors in a rod probe, for example. However, it must be viewed critically that basic requirements for sensor installa-

3

Table 3.23 Recommendations for different installation variants

Application	Temperature	Soil Heat flux	Relevance
a	Side wall of a profile pit		
b	Rod probe or Side wall of a profile pit		
c	Side wall of a profile pit	Heat flux plate or temperature gradient (side wall of a profile pit)	
d	Rod probe or Side wall of a profile pit	Heat flux plate (side wall of a profile pit)	

tion can often only be partially realized. Soil compaction at the borehole and potential heat and water conduction along the sensor can lead to incorrect readings. Table 3.23 gives recommendations for different installation variants.

3.8.3 Which Method is to Be Used for Measurement?

While the soil temperatures can be measured analogous to ▶ Sect. 3.1, the soil heat flux is measured directly as well as determined from soil temperature and moisture measurements.

Temperature measurements are required to determine the soil temperature and the heat storage in the soil above the soil heat flow measurement. Waterproof moulded sensors, mainly Pt100 but also thermocouples, are used for this purpose. For details see ▶ Sect. 3.1.3.

The principal schema of a ***heat flux plate*** is shown in Fig. 3.18. A row of thermocouples (Fig. 3.4) is connected to the top and bottom of the plate. The sensor body has an averaged thermal conductivity of the soil. Errors due to different thermal conductivities can be reduced by the Philip correction (Philip 1961), which should mainly be applied to non-self-calibrating heat flow plates (Liebethal and Foken 2006). Therefore, the soil heat flux can be calculated from the temperature difference between the top and bottom surfaces. Self-calibrated plates have a heater on the top surface. The heater generates a known heat flux for a short period of time. From this flux, the defect error can be calculated, and the data can be corrected. This type of plate has the advantage of higher accuracy but has a significant disadvantage in that the data cannot be used during and immediately after heating.

For meteorological measurements, the ***ground heat flux*** at the surface is usually required rather than at a specific depth. In addition, the temporal change in the storage term above the ground heat flux plate must be determined by means of additional temperature measurements. The ground heat flux is then the sum of the change in heat storage above the heat flux plate and the heat flux measured with the plate (Liebethal and Foken 2007; Yang and Wang 2008; Gao et al. 2017; Huwe et al. 2021). As an alternative to the soil heat flux plate, the heat flux at a certain depth can be determined by the temperature gradient (e.g. for a depth of 20 cm, take the gradient between 15 and 25 cm) using a_G, the molecular heat conduction coefficient (Foken et al. 2021):

$$Q_G(-z) = a_G \left. \frac{\partial T}{\partial z} \right|_{-z}. \tag{3.12}$$

The calculation of the change in heat storage above the soil heat flux measurement is shown in Note 3.7.

Note 3.7

Calculation of the change in heat storage in the soil

The easiest way to determine the storage term is to measure a layer medium temperature with an integrating temperature sensor (measuring element measures over the entire sensor length) over the entire soil layer between the soil heat flux measurement and close to the surface (Liebethal et al. 2005; Bonan 2019; Huwe et al. 2021; Foken and Mauder 2024), so that (not recommended for application c)

$$Q_G(0) = Q_G(-z) + \frac{C_G \cdot |\Delta z| \cdot \left[\overline{T(t_2)} - \overline{T(t_1)}\right]}{t_2 - t_1}. \quad \text{(N3.6)}$$

It is more complex to measure several temperatures above the floor heat flow plate and determine the change in storage over time according to

$$Q_G(0) = Q_G(-z) + \int_{-z}^{0} \frac{\partial}{\partial t}[C_G(t,z)\ T(t,z)]dz. \quad \text{(N3.7)}$$

The volumetric heat capacity required to determine the storage term can be determined using a method developed by de Vries (1963)

$$C_G = C_{G,m}x_m + C_{G,o}x_o + C_{G,w}\theta, \quad \text{(N3.8)}$$

with the heat capacities for mineral and organic components ($C_{G,m}$=1.9 × 10^6 J m^{-3} K^{-1}, $C_{G,o}$=2.479 × 10^6 J m^{-3} K^{-1}) and for water ($C_{G,w}$=4.12 × 10^6 J m^{-3} K^{-1}), whereby x_m is determined as the proportion of mineral components (assumed mineral density of 2650 kg m^{-3}) from volume measurements of the soil and x_o can often be neglected for a depth layer of up to 20 cm. The volumetric moisture of the soil is θ and, like x_m, is given in m^3 m^{-3}. This requires either integrating soil moisture measurement methods or measurements at different depths above the soil heat flux plate.

3.9 Soil Moisture and Soil Water Tension

Historically soil water content was measured simply by the gravimetric method. One of the first accurate techniques to monitor volumetric soil moisture was the Neutron moderation method, developed in the 1950s. As the equipment needed was not only very expensive but also required a very skilled operator authorized to handle nuclear material, it was necessary to serve as many clients and take measurements in as many fields as possible with one instrument. Thus, usually only one reading could be obtained per week, with the gaps having to be filled by interpolation methods.

In recent years, many companies have introduced a wide range of various soil moisture sensors, which are now available for just about any budget, many applications, and in many different technologies. There is no ideal sensor, one that does it all, one that is equally suited for sand and clay, for blueberries and grapes, turf, and pecans. But within the range of available products, everyone will be able to find one that meets the criteria for the task at hand.

3.9.1 Why Should Soil Moisture Be Measured?

From a micrometeorological point of view, soil water content strongly influences the latent heat flux above ground or vegetated surfaces and therefore also the sensitive heat flux part of the energy balance equation as well as the Bowen ratio (see ▶ Sect. 1.1.4). The ground heat flux is strongly determined by soil water content. The course of soil water content over the year is an important indirect indicator of seasonal (micro)climatic and climatic characteristics (see ▶ Sect. 1.1.5). Indirectly, by influencing the Bowen ratio (sensitive vs. latent heat flux) it is considerable influencing the variability of vegetation canopy conditions and canopy and near-ground microclimatic variables, such as surface and canopy air temperatures and air humidity.

The idea behind direct soil moisture measurements or its monitoring for practical applications is simple to reduce uncertainties from indirect modelling assessments. For example, instead of determining the amount of irrigation by modelling the soil-crop water balance where especially the soil model input data as well as the actual rooting zone of a crop are often unknown or uncertain due to high spatial variability in soil properties (▶ Sect. 2.3.1), the direct measurement of the actual soil water status at the plant's active root zone at the representative site can in-

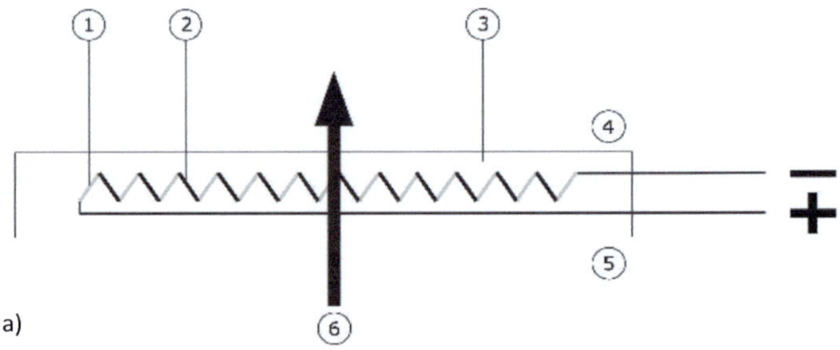

a)

b)

Fig. 3.18 **a** Principal schema of a heat flux plate with a number of thermocouples of two different alloys, where the connections of the alloys are connected to the upper and lower sides of the plate. (1: alloy 1; 2: alloy 2; 3: sensor body with an average thermal conductivity; 4: upper sensor surface; 5: lower sensor surface; 6: soil heat flux); **b** Heat flux plate (image used with permission from Hukseflux, Delft, The Netherlands)

Table 3.24 Application purpose for soil moisture measurements

Code	Application	Relevance
a	Input variable for irrigation scheduling (crops, trees, etc.)	
b	Forecasting field conditions for crop management options	
c	Analysis of microclimatic canopy conditions affecting pest, and disease risks	
d	Input variable for local climate models	
e	Calibration and validation parameter for agronomic models (crop model) and hydrologic models	
f	Control parameter for flux estimations with the → eddy-covariance method	
g	Control parameter for remote sensing products	

crease the assessment quality of actual crop water demand significantly.

Soil moisture sensors, however, offer much more insight into what is happening in the soil and the root zone than just measuring soil moisture. They indicate the infiltration rates of irrigation and natural precipitation, they allow us to assess water table fluctuations, they assist in determining agronomical thresholds such as refill point/wilting point and field capacity, and last but not the least they help us to properly determine the dimension of the active root zone and its variability over time in the various phenological phases. As is known, soil water content is also an indicator of agricultural drought. Soil water content as an indirect indicator of crop canopy humidity conditions is affecting conditions for fungi diseases and risks for inoculation. Further it is an indicator for soil stability against field traffic by machinery (e.g. for soil cultivation, sowing, fertilizing, and harvesting), where soil compaction risks can be estimated or avoided if the actual conditions are known or forecasted. Typical application examples for soil moisture measurements are given in ◘ Table 3.24.

3.9.2 Where and How to Measure Soil Water Content or Tension?

Under natural conditions, soil water content can develop strong vertical gradients in the vertical soil column over growing seasons with high variability and dynamics depending on precipitation/irrigation events, or ground water-level impact. Soil water content is influenced further by the often highly variable soil properties in different vertical soil layers as well as in the horizontal space, which are mostly more or less static, but can also strongly change due to soil disturbances such as soil cultivation at the topsoil layer. Plant roots extract considerable amount of water from the actual rooted zone, adding variability depending on vegetation status, season, or weather conditions.

Considering above-mentioned influencing factors, sensors for measuring soil water content or soil water tension should be carefully selected as well as sited in order to receive representative measurements according to the measurement task. For example, for receiving representative or statistical valid measurement results under natural conditions of varying soil properties at the measurement site, two main aspects, except the technical suitability, are relevant:

- The size and type of the sensor probes, in order to measure a large soil volume (larger measured volume will reduce the small-scale impacts of disturbing elements such as holes and gaps, stones, and others in the soil).
- Number of replications (at least 3 replications are recommended for the same siting condition (e.g. specific soil layer or different soil condition/type in the horizontal scale). Replications are also important to avoid data gaps in case of sensor failure, which can never be ruled out.

Task-oriented siting of sensors in the soil include siting sensors within the rooting zone of crops for estimating crop water demand, drought status, or any soil-crop water demand. In another case of calibration/validation purposes for remote sensing methods, which can measure directly only a small soil top layer, a high number of spatial replications of the topsoil layer is necessary (◘ Table 3.25).

According to the lengths of sensor rods (as usual for FDR or TDR/TDT systems, see ▶ Sect. 3.9.3) of a certain single sensor, there is the option to place a sensor in the vertical in the soil (measuring a vertical integral over a certain soil depth, for example 0–20 cm soil depth) or in horizontal over a vertical soil profile, addressing better characteristics of different soil layers. In regard to measurement of vertical soil water profiles, it can be recommended to increase the sensor distance with soil depth, as the variability and dynamics of soil water content is decreasing

◘ **Table 3.25** Measurement siting for soil water measurements according to tasks

Application	Measurement siting	Relevance
a, e	According to the plants rooting depth, multiple measurement depths at the side wall of a profile pit or using a vertical tube design	
d, f, g	Topsoil layer, many replications, rod probes	
b, c, e	Topsoil layer, few replications with representative siting (e.g. below crop canopy), rod probes	
e, f, g	Deep soil columns, soil layer oriented (hydrological aspects) at the side wall of a profile pit or using a vertical tube design	

with soil depth, for example 10–20–40–80 cm depth steps under relatively homogeneous soil conditions (strong differences in soil properties between certain soil layers should be considered separately).

3.9.3 Which Method is to Be Used for Measurement?

The soil water content is measured simply by the gravimetric method, applying scaling before and after drying of a soil probe. This method, which is still in use as a general control parameter, however, provides only a rough assessment of plant available water and hardly allows to produce longer time series of data due to the workload of soil probing. In order to receive the volumetric soil water content from the gravimetric method, it needs the probing by soil cylinders (for defining the reference soil volume, the gravimetric estimated soil water content and the estimated soil bulk density after drying); see Lalic et al. (2018).

Recent technical developments, however, have brought the advance of less costly and more user-friendly equipment (see below). In conjunction with a data logger, it has now become affordable to leave an instrument permanently on site and to take a reading, for example every 15–30 min which is suitable for most applications. This means that if the logger is also connected to a telemetry device, data can then be sent to the meteorological service or to the farm manager in almost real time. In the case of needing to adjust one's daily irrigation practice flexibly to changing conditions, a permanent monitoring system in combination with a telemetry device is clearly the preferred option. For practical field operations in agriculture, permanent monitoring significantly reduces the time needed to adapt irrigation needs. Furthermore, daily observation of plant water uptake from the actual root zone is very helpful to better understand its varying physiology during the different phenological phases.

Several measurement methods can be distinguished in order to determine soil moisture levels which are the spatial scale addressed ones—single- and multi-level sensing—and the time scale addressed ones—the frequencies at which readings are taken.

Soil moisture conditions with in situ sensors can be basically determined by

- measurement of volumetric soil water content (various methods available) and
- measurement of soil water tension (or potential).

Most sensors used these days employ some form of volumetric methods (◘ Fig. 3.19). They give a direct reading of soil water status (units of water per unit of soil, e.g. cm^3 of water/cm^3 of soil, or simply expressed in %Vol.), but shed no light on the ability of a plant to extract water from the soil (which needs the measurement of soil water tension/potential). Many soil water sensors are nowadays already combined with integrated soil temperature sensors (◘ Fig. 3.19a–d).

The technology to measure volumetric soil water content is based on determining the dielectric constant of soil, indicating the capacity of a non-conductor (i.e. soil as isolator) to transmit an electromagnetic wave, such as the one emitted by a soil moisture sensor. As the dielectric constant of dry soil and that of water are known and strongly different, a curve can be developed correlating the sensors' response to the moisture content of the soil, given that this content is very low in dry soil and very high in saturated soil. Two main types of in situ measurement techniques are used nowadays (Robinson et al. 2003; 2008; Huwe et al. 2021):

- Frequency domain reflectometry (FDR)/capacity-based;
- Time domain reflectometry/transmissivity (TDR/TDT)-based.

The capacitance or FDR technique determines the dielectric permittivity of a medium by measuring the charge time of a capacitor, which uses that medium as a dielectric. In contrast, time domain reflectometry/transmissivity (TDR/TDT) determines the dielectric permittivity of a medium by measuring the time it takes for an electromagnetic wave to propagate along a transmission line that is surrounded by the medium. TDR measurements are theoretically less susceptible to soil and environmental conditions compared to capacitance sensors. However, the interpretation of TDR output can be a considerable source of error when, for example, high salinity diminishes the reflectance waveform.

Rapidly advancing sensor development has led to the multi-level probing (measuring vertical soil profiles), developed in the early 1990s, which allows the comfortable installation of several sensors on one site without the need to disturb the root zone by digging a profile (see ◘ Fig. 3.19a). Such sensors are usually installed through an access tube which is installed into the soil, and can be spaced at various sensor intervals, up to a placement density of one sensor every 10 cm. They can deliver accurate readings right after installation, without the need to wait for several weeks or months until the soil has settled again from digging in several individual sensors. However, since multi-level sensors of the access-tube kind necessarily determine soil water content without being in direct contact with the medium, they are also subject to some limitations. Pricewise, single-level sensors are less expensive than multi-level sensors and are usually installed for crops with shallow roots or for applications such as irrigation scheduling.

For application on larger scales (measuring footprint of about 100–200 m), a relatively new (still under development) and non-invasive promising technique is the CRNS (cosmic-ray neutron sensing) which is filling the gap between the small-scale information provided by point soil moisture sensors and the large-scale data provided by remote sensing. This method is integrating over the measured path and cannot resolve small-scale variations of soil water content, the measuring footprint is decreasing with increasing soil moisture, and it needs to be calibrated (see for further reading, e.g. Jakobi et al. 2021).

For specific application, e.g. long-term monitoring in remote places, specific autarkic robust sensor types were developed recently (◘ Fig. 3.19d), allowing long-term measurements with up to 10 years data storage, and battery lifetime. However, these sensor data are not transmittable online due to energy demand so far.

Soil water potential or tension is measured by the matrix potential or soil suction method often referring to as the force a plant needs to extract water from the soil. The unit of measure is kilopascals. Such sensors are rather inexpensive by comparison, but require more attention and maintenance than volumetric sensors. Their common denominator is the use of a porous media such gypsum or specific designed materials that allows water to penetrate into the sensing volume for the water content depending electric conductivity. Examples of well known sensor names are tensiometers, gypsum blocks, and Watermark sensors (◘ Fig. 3.20). These sensors differ, however, in the measurable range on soil water potential, and need, especially for measuring absolute soil water potentials, regular calibration to the relevant soil types. Depending on the material used, the lifetime of these sensors is limited, especially for the low-cost gypsum blocks.

3.10 Combination Sensors

In recent years, more and more combined sensors have been offered that combine the measurement of several atmospheric variables (temperature, humidity, wind, radiation, and even precipitation) in one device. These devices are less expensive compared to a single sensor set of a similar quality, but a certain amount of interferences (e.g. wind field is disturbed by the compact design) cannot be completely excluded (see ◘ Fig. 3.21). They are used wherever very high accuracy is not required (Wichura and Foken 2021), e.g.:

- Weather monitoring for rural (e.g. crop management) applications and in urban locations such as street canyons, and parks.
- Measurement of climate elements that are relevant to spa conditions.
- Measurement of variables relevant to renewable energy (wind, solar, water).

3

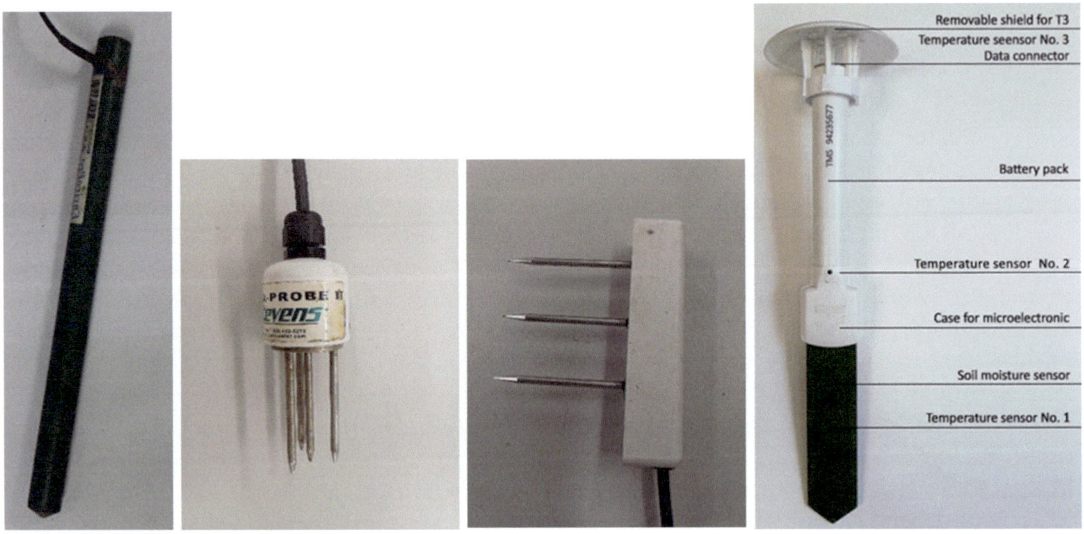

a) b) c) d)

Fig. 3.19 Exemplary volumetric soil moisture (incl. soil temperature) probes of different designs and application potentials from different producers. From left to right: **a** EnviroPro sensor (multiple soil depth capacitance sensors in plastic tube of different lengths), Entelechy Pty Ltd., **b** Hydra Probe (TDR), Stevens Water Monitoring Systems, **c** TEROS 11 sensor (FDR/capacitance-based), METER Group (Photos a–c: Eitzinger, BOKU), and **d** TOMST (autarkic sensor, TDT). *Source* TOMST s.r.o.

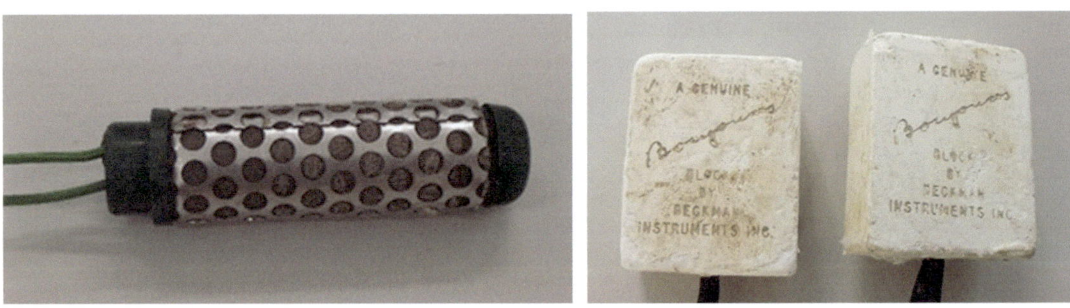

Fig. 3.20 Soil water potential sensors for agrometeorological use: **a** Watermark sensor (special granular matrix), Irrometer Company Inc.; **b** Gypsum blocks, Beckman Instruments Inc. (all photos: Eitzinger, BOKU)

- Determination of meteorological elements that influence sound propagation.
- Measurement of meteorological elements that may affect traffic.

For the meteorological elements listed in Sects. 3.1–3.6, the combined sensors can be used for the applications listed in Table 3.26.

There are also sensors that measure soil temperature and soil moisture together in a vertical profile (Fig. 3.21) and are used, e.g. for forest fire warning, but also for applications mentioned in Sects. 3.8 and 3.9; see Table 3.26.

3.11 Evapotranspiration Measurements and Calculations

Evapotranspiration (ET) describes evaporation as a combination of transpiration (T) from living vegetation (i.e. evaporation through leaf stomata and epidermis) and di-

a) b)

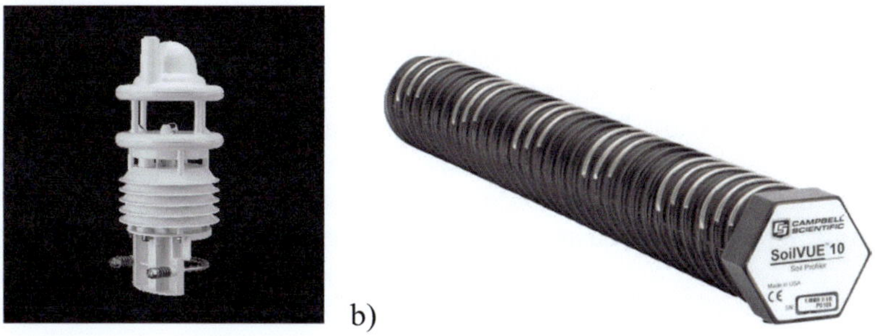

Fig. 3.21 Combined sensors for **a** meteorological elements (Photo: OTT Hydromet); **b** soil temperature and moisture (Photo: Campbell Sci.)

Table 3.26 Possible applications of combined sensors

Meteorological element	Table	Application	Relevance
Air temperature	3.1	c, d, e, (f)	
Air humidity	3.6	c, d, (e)	
Wind speed and direction	3.9	c, d, (e)	
Atmospheric pressure	3.13	a, b	
Global radiation	3.16	c, (d)	
Precipitation	3.19	b, (c)	
Dew fall	3.21	b, d	
Soil temperature	3.22	b, d	
Soil moisture	3.24	a, b, c	

rect evaporation (E), which is the evaporation from all uncovered surfaces, and open water bodies. Interception as part of evaporation is the evaporation of wet leaves after rain or dewfall (see ► Sect. 3.7). Sublimation takes place when the surface is frozen. Potential evaporation is the evaporation from water-saturated surfaces and water surfaces.

In the following, the direct measurement of evapotranspiration using lysimeters and potential evaporation using evaporation boilers (evaporimeters, evaporometers, or atmometers) are described, along with a number of other methods, as well as calculation procedures. Methods for determining evapotranspiration based on the measurement of turbulent atmospheric exchange are dealt with in ► Sect. 3.12.

The first really usable evaporometer was built by Albert Piche (1872) (Atmometer), who sealed a glass tube filled with water with a filter paper and measured the water loss. A large number of similar devices have been developed to date, which measure the evaporation of water-saturated ceramic plates, among other things. It is difficult to transfer these measurements to natural conditions of a larger spatial scale, especially as ambient conditions, wind speed, and radiation have a significant influence on the single point measurements. On the other side, they can be used to distinguish the effect of small-scale microclimates to atmospheric evaporative demand.

A lysimeter measures not only the actual evapotranspiration but also the percolation processes and the utilization of water by the vegetation and are since already about 300 years in use (Reth et al. 2021) with many different designs and techniques.

Simple empirical evaporation calculation methods were originally based on the saturation deficit according to Dalton (Foken and Mauder 2024) or air temperature (Turc 1961). Their use must be questioned today, as climate change means that the conditions under which the relationships were determined no longer apply. Today, relationships based on the energy balance equation or on resistance approaches are used, with Penman (1948) method for potential evaporation and Penman-Monteith method (Monteith 1965) for actual evaporation forming the basis.

3.11.1 Why Evapotranspiration Should Be Determined?

Evapotranspiration measurements and calculations are needed for various purposes in meteorology, climatology, hydrology, and agronomy. Evapotranspiration for micrometeorologists appears as one of the most important terms of the surface energy budget as the latent heat flux. Its dimension is W m^{-2} (see Note 1.1), while in hydrological and agricultural applications, e.g. irrigation planning, it is given in mm, like precipitation. The goal of investigation determines which measuring and modelling technique and accuracy and which time scale (e.g. from half hourly to daily variation of energy budget components to monthly scale estimations for climatology and planning) is needed. Typical application examples are given in ◘ Table 3.27.

3.11.2 Where to Measure?

Field- and regional-scale actual evaporation calculation methods are distinguished. Regional evaporation is given for an area larger than 1 km^2 (for example a farm, small catchment with different land-cover types) usually on daily, weekly, and monthly scales. This methodology also supports solving farm and catchment basin-level water management tasks. Therefore, direct measurements as well as measurements of the input data for modelling approaches should be measured in such a way that they are as representative as possible for the area of interest. Depending on the model input data, the corresponding sections must be taken into account.

For direct measurements (▶ Sect. 3.11.3), it is essential that the environment of the measuring device fulfils the required conditions or is completely uniform in terms of vegetation and soil properties. A particular problem arises if the surroundings of the measuring device are drier. Then drier air flows over the measuring device with the wind field, the vertical gradient of the vapour pressure increases, and thus a higher evaporation or evapotranspiration is simulated. The conditions correspond to the oasis effect (see ▶ Sect. 1.2.7). For measurements to determine evaporation with model approaches, uniform surface properties are required over the entire footprint area (see ▶ Sect. 1.2.4).

3.11.3 Which Method is to Be Used for Measurement?

Both direct measurements for potential evaporation and actual evapotranspiration as well as modelling approaches for potential evaporation and actual evapotranspiration are presented below (Allen et al. 2021a). The focus

◘ **Table 3.27** Application purpose for evapotranspiration measurements and calculations

Code	Application	Relevance
a	Surface energy and water budget estimation in different time scales	
b	Climate analysis, local climate	
c	Characterization of human well-being, climate resilient-planning (Vulova et al. 2023)	
d	Determination of evapotranspiration-dependent climate indices → e.g. aridity indices	
e	Irrigation and drought monitoring	
f	Forecasting field conditions for crop management	
g	Agronomical and hydrological planning on different scales	

is on the presentation of important principles. There are many modifications, most of which only apply to a specific area under specific climatic conditions (WMO 2020; Foken and Mauder 2024). If you only need an estimate of evapotranspiration, you can also use the model data provided by some meteorological services on a grid element basis.

Evaporimeters, evaporometers or atmometers are synonyme names of devices which measure water evaporation from an open water body or the amount of water evaporated through a porous medium such as ceramic plate or sphere (◘ Fig. 3.22). In follows we call all devices "Evaporimeter".

A classic standardized hydrological method for determining potential evaporation is the evaporation pan, mainly the ***Class A Pan*** (◘ Fig. 3.22, DeFelice 1998). This is a round pan with a water surface area of 1.14 m^2 (diameter 120.65 cm, height 25.4 cm) and a water depth of 15.2–17.8 cm. The evaporation is measured by the water loss in the pan—corrected with the precipitation—and modified depending on the wind speed, humidity, and water temperature. A correction factor for the grass reference evapotranspiration (Allen et al. 1998) or other wet reference surfaces has to be calibrated and applied, which is in the range of 0.6–0.8 in most cases. A highly simplified but reliable method for daily sums of evaporation in mm d^{-1} (Smajstrla et al. 2000) uses the height difference of the water level in mm in the evaporation pan and corrects this with a factor given in ◘ Table 3.28.

Evaporimeters are relatively simple and cost-efficient devices to measure evaporation. Besides the standardized Class A pan, the use of small (not standardized) pots with open water surface where evaporation is estimated by weighting differences between selected time periods allows a higher flexibility in selecting measurement sites, to detect the atmospheric evaporative demand at very small-scale microclimates (e.g. inside and outside of a plant canopy). Similarly, evaporimeters of different designs with permeable ceramic evaporation surfaces (where only distilled water can be used!) are suitable for analogue

a) b)

◘ **Fig. 3.22** **a** Class-A-pan evaporimeter with a cylinder to measure the water level; temperature and wind sensors are also present (photo: Hong Kong Observatory); **b** evaporimeters of different designs (1—manual pot evaporimeters; 1a—scale; 2—manual ceramic evaporimeter; 3—electronic ceramic evaporimeter with green cup mimicking albedo of vegetation (ET Gage, Photo: Eitzinger, BOKU))

Table 3.28 Coefficient for the determination of the evaporation with the Class-A-Pan for meadows or grain (in brackets for bare soil) without water shortage in the vicinity of the device (Doorenbos and Pruitt 1977; Smajstrla et al. 2000)

Mean wind velocity at the measurement date	Extension of the area in the surrounding in m	Minimum of the relative humidity at the measurement date	
		>40%	<40%
Low ≤2 ms^{-1}	1	0.65 (0.80)	0.75 (0.85)
	10	0.75 (0.70)	0.85 (0.80)
	100	0.80 (0.65)	0.85 (0.75)
	1000	0.85 (0.60)	0.85 (0.70)
Moderate 2.1–4.4 ms^{-1}	1	0.60 (0.75)	0.65 (0.80)
	10	0.70 (0.65)	0.75 (0.70)
	100	0.75 (0.60)	0.80 (0.65)
	1000	0.80 (0.55)	0.80 (0.60)
Strong ≥4.5 ms^{-1}	1	0.50 (0.65)	0.60 (0.70)
	10	0.60 (0.55)	0.65 (0.65)
	100	0.65 (0.50)	0.70 (0.60)
	1000	0.70 (0.45)	0.75 (0.55)

or even automated measurements. These devices, however, do not deliver high absolute accuracy if not calibrated against a relevant standard method, as the measurements are affected by the specific technical design and materials affecting evaporation resistances. However, they are still very suitable for relative comparisons of evaporation affected by modified microclimate at small spatial scales, such as those applied also in transect measurements (see example in ▶ Sect. 2.7.3).

Evapotranspiration is the evaporation and/or transpiration from surfaces (e.g. bare soil or vegetation) with occurrence of water shortage, where different bulk resistances can limit evapotranspiration (e.g. stomata closure or dry soil surface). For such conditions, ***lysimeters*** of different designs are often used for evapotranspiration measurements in agricultural meteorology, which are a good but—depending of its size—labour- and cost-intensive direct measurement technique. The advantage is that, in contrast to micrometeorological measurements, only small areas are required. With this method, a vegetated soil monolith is inserted into the ground and weighed (alternatively soil water change is measured by sensors in not-weighted lysimeters which is however less accurate for evapotranspiration estimate), while simultaneously measuring precipitation and water drainage out of the soil column (▫ Fig. 3.23a, Reth et al. 2021). Soil temperature and soil moisture in the lysimeter should represent the area conditions to be measured, i.e. largely be identical in their vertical structure. These measurements often show good agreement with eddy-covariance measurements (see ▶ Sect. 3.12), when corrected for the unclosed energy balance (Mauder et al. 2018), e.g. by an oasis effect due to an unsuitable fetch (see ▶ Sect. 1.2.4). In general, lysimeter measurements can be affected by several further disturbances and uncertainties which need to be carefully checked and corrected (e.g. for the wind pressure affecting the scale, depending on crop height). Mini lysimeters are a simplified, small, cost-efficient, and easy-to-use form which are often placed above the ground or used in laboratory trials (▫ Fig. 3.23b). Their big advantage is that they are mobile and flexible in installation and the user doesn't need to necessarily consider the fetch when applied for estimating microscale conditions. Their disadvantage is that the specific design can considerably modify evaporation conditions (e.g. placement above the ground) in comparison to surrounding natural conditions.

A commonly used method for the ***calculation of the potential evaporation*** was proposed by Penman (1948). The evaporation is calculated from the available energy (radiation balance reduced by the ground heat flux) and the Bowen ratio as well as a wind speed-dependent correction term (Foken and Mauder 2024). The ***Priestley-Taylor method*** (Priestley and Taylor 1972) used as an intermediate step in the derivation is less complex (Note 3.8). However, the Priestley-Taylor coefficient varies from region to region and may need to be adjusted (Szilagyi et al. 2015). Input data must be provided on a half-hourly or hourly basis, but there are derivates also for daily time steps used for e.g. ecosystem, crop and agrometeorological models. However, the results can only be used with sufficient accuracy for decade or monthly values.

Note 3.8

Priestley-Taylor approach (hourly time step):

From the Priestley-Taylor approach (Priestley and Taylor 1972) follows for the sensible heat flux

$$Q_H = \frac{\left[(1-\alpha_{PT})\, s_c + \gamma\right]\left(-Q_s^* - Q_G\right)}{s_c + \gamma} \tag{N3.9}$$

and for the latent heat flux (evapotranspiration)

$$Q_E = \alpha_{PT}\, s_c\, \frac{-Q_s^* - Q_G}{s_c + \gamma}, \tag{N3.10}$$

with the Priestley-Taylor coefficient $\alpha_{PT} \sim 1.25$ (typical value for water surfaces), the net radiation $-Q_s^*$ ($-Q_s^* > 0$ at daytime), the ground heat flux Q_G, the psychrometric constant γ, and the change of the saturation vapour pressure with temperature s_c. Typical values of γ and s_c are given in table (Stull 1988):

Temperature in K	γ in K^{-1}	s_c in K^{-1}
270	0.00040	0.00022
280	0.00040	0.00042
290	0.00040	0.00078
300	0.00041	0.00132

The transition from the Penman to the ***Penman-Monteith approach*** (Penman 1948; Monteith 1965; DeBruin and Holtslag 1982) included the consideration of non-saturated surfaces and cooling due to evaporation, which reduces the energy of the sensible heat flux. Including both aspects leads to the Penman-Monteith method for the determination of the actual evaporation (evapotranspiration). The ventilation term determining the cooling is calculated from the stomatal resistance and the turbulent atmospheric resistance (Foken and Mauder 2024). The method is clearly defined in physical terms and does not contain any empirical coefficients. There are a large number of similar methods (Shuttleworth and Wallace 1985; Szilagyi and Crago 2023), but when applying them, it must be checked whether they are suitable for the area of application and whether modifications may be necessary due to climate change.

For facilitating practical applications in agriculture (e.g. crop irrgation estimation), the Food and Agriculture Organisation of the United Nations (FAO) uses the Penman-Monteith method for comparative analysis of evapotranspiration—***FAO grass reference evaporation*** (Allen et al. 1998; Moene and van Dam 2014), calculated at daily time steps, which considers the global availabilty of data from weather stations. A fixed set of parameters was defined for this purpose (Note 3.9). For application at any location, the parameters would have to be changed accordingly. In drought, the FAO method overdetermines the evapotranspiration (Allen et al. 2021b) and the stomatal resistance would have to be increased (Pirasteh-Anosheh et al. 2016).

The idea to define a reference evapotranspiration obviously goes back to Doorenbos

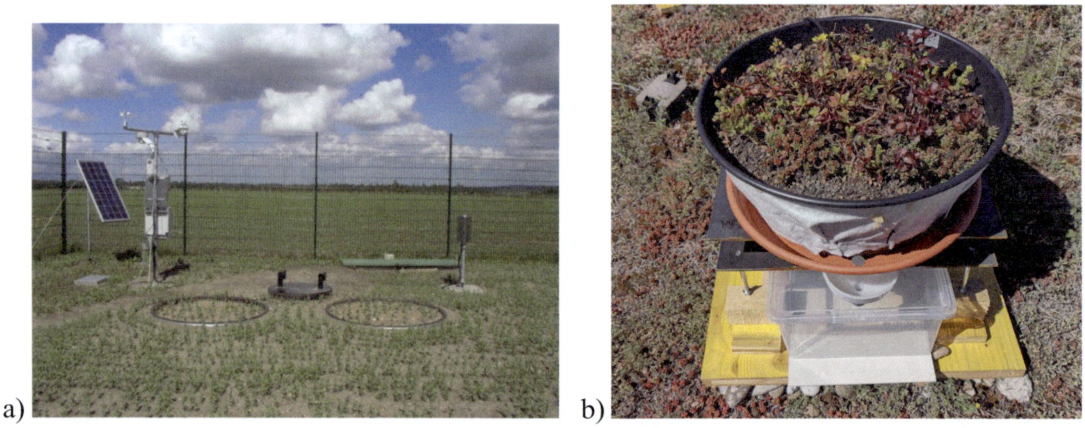

Fig. 3.23 **a** Double lysimeter, © Umweltgerätetechnik GmbH, Müncheberg, Germany, **b** Mini lysimeter applied for green roof vegetation (electronic weighting lysimeter including percolation collector), own design, BOKU, Vienna, Austria. Photo Eitzinger, BOKU

and Pruitt (1977), who also included a so-called crop coefficient K_c, which varies depending on plant species, plant height, and climate. A comprehensive overview of different crop coefficient assumptions was given by Allen et al. (2021a).

Note 3.9

FAO-grass-reference evapotranspiration: The evapotranspiration according to the FAO (Allen et al. 1998) at daily time step is given by

$$Q_E = \frac{s_c\left(-Q_s^* - Q_G\right) + \rho\, c_p \frac{0{,}622}{p}\, \frac{E-e}{r_a}}{s_c + \gamma\left(1 + \frac{r_s}{r_a}\right)}, \tag{N3.11}$$

with the net radiation $-Q_s^*$, the ground heat flux Q_G, the psychrometric constant γ, the change of the saturation vapour pressure with temperature s_c (typical values of γ and s_c are given in Note 3.8), the saturation vapour pressure E, the vapour pressure e, the turbulent atmospheric resistance r_a=208 u^{-1}, wind speed at 2 m height in m s^{-1}, and the stomata resistance r_s=70 s m^{-1}. These values are based on a canopy height of z_B=0.12 m and a roughness length of 0.123 z_B.

3.12 Flux Measurements

The determination of the turbulent fluxes of sensible heat, latent heat (evapotranspiration), and trace gas fluxes as well as the momentum flux are fundamental methods in micrometeorology and increasingly also in forestry and agricultural meteorology. There is no special device for this, but the measurements are realized by combining various suitable devices, as already presented in this chapter. The measurements are extremely complex and often require specialized knowledge, so that only an overview can be given here. Reference is made to the specialized literature. It is generally recommended to take advantage of training courses and field practical or to consult experts before starting these measurements.

Measurements have been carried out for many decades. Initially, vertical profiles of wind speed, air temperature (ventilated), and humidity were measured at around 4–6 measurement heights at logarithmic intervals in the lowest approximately 10 m. Such measurements were very time-consuming to carry out and to correct and were limited to large-scale flat terrain due to the footprint and internal boundary layers. At present, methods with two measuring heights are still used, whereby the upper measuring height is a maximum of 3–4 m (above water up to 10 m) and the lower measuring height is at about 2 times the height of the canopy or at the water surface. Due to the logarithmic profiles of wind speed, air temperature and air humidity and the modification of the profiles by atmospheric stratification, the evaluation is not trivial here either. Only the Bowen ratio method (see ▶ Sect. 3.12.3) is relatively easy to apply. Foken and Mauder (2024) provide an overview of these methods, so that further explanations are not necessary.

Profile measurements are rarely used today. They have largely been replaced by the eddy-covariance method. The turbulent fluctuations of the vertical wind, horizontal wind, and scalars are measured directly, and the flux is determined from this. The method was first used when sonic anemometers (see ▶ Sect. 3.3.3) became available in the 1960s. In the last 30 years, the method has become popular, particularly through the FLUXNET programme for determining the fluxes of water vapour and carbon dioxide (Baldocchi et al. 2001; Schmid and Rebmann 2021), and was no longer limited to a narrow circle of experienced experts. Comprehensive literature and data processing software are now also available (Aubinet et al. 2012; Mauder et al. 2021). Beginners are referred to the book by Burba (2022). Training or guidance from experts is essential for the application of the method. Some basics are given in ▶ Sect. 3.12.3.

3.12.1 Why Should Turbulent Fluxes Be Measured?

Today, flux measurements are an integral part of many complex measurement programmes on the energy and matter exchange of ecosystems and, increasingly, cities. The measurements are always carried out in combination with classic climate measurements and, when recording the complete energy balance at the earth's surface, also net radiation measurements and soil heat flux measurements. Possible areas of application are listed in ◘ Table 3.29. However, the Bowen ratio method cannot be used over high vegetation, because the gradients are too small there and the error of the measurement is very large.

While the Bowen ratio method and the eddy-covariance method are point measurements and the result must be assigned to the corresponding footprint, scintillometer measurements are area-averaged along a measuring path, taking into account the footprint along this measuring path (Leclerc and Foken 2014).

3.12.2 Where to Measure Turbulent Fluxes?

The feasibility of flux measurements depends largely on the environmental conditions. In addition to a largely flat terrain with uniform vegetation (uniform footprint), this includes the need to ensure that no internal boundary layers or obstacles can influence the measurements. For Bowen ratio measurements, it is important that both measurement heights are assigned to the same footprint area and are located in the new equilibrium layer of a possible internal boundary layer.

The measurement heights for the Bowen ratio method apply above the zero-plan displacement. The distance between the lowest measurement height and the canopy height should be as low as possible, but at least 0.2 m, and must also be maintained if the stand is growing, and correspondingly greater above tall canopies. Eddy-covariance and scintillometer measurements should be taken at least 3–4 m above zero-plane displacement. Although the flux measurement is independent of the height, it must be taken into account when making the necessary corrections (see software documentation). See ▶ Chap. 2; which atmospheric factors above the surface to be analysed must be taken into account. The advice of an expert is absolutely necessary here.

3.12.3 Which Method is to Be Used for Flux Measurement?

A large number of methods for flux measurements are described in the literature (Foken and Mauder 2024). Many of these methods are very complex and, apart from the techni-

3

Table 3.29 Applications for flux measurements

Application	Method			Description	Relevance
	Bowen ratio	Eddy-covariance	Scintillometer		
a	x	x	x	Energy balance at the earth's surface	
b	x	x	x[b]	Evapotranspiration	
c		x		Carbon dioxide flux	
d	x[a]	x		Trace gas fluxes	

[a] Modified Bowen ratio method; [b] Microwave scintillometer

cal complexity of the equipment, cannot be used without extensive specialist knowledge. Beginners should focus on the Bowen ratio method, which is described in more detail below. It is certainly the wish of many users to use the modern eddy-covariance method. Only a brief introduction is given and, as mentioned above, the user should first acquire the necessary knowledge by attending appropriate training courses. A combination of both methods is the modified Bowen ratio method, which can be used to determine trace gas fluxes and is also briefly described below. Measurements such as those carried out with the Bowen ratio method can also be used to model trace gas fluxes (Trebs et al. 2021).

One more preliminary remark is important: The Bowen ratio method is based on closing the energy balance and distributes the excess energy from the sum of the net radiation and the ground heat flux to the sensible and latent heat flux. In contrast, the eddy-covariance method is based on the equations of atmospheric turbulence without forcing the closure of the energy balance. Especially in heterogeneous terrain, there are larger turbulence structures that transport energy but cannot be determined by local measurements. Under these circumstances, the measured fluxes are lower and do not close the energy balance. There are correction methods to determine the proportion of larger turbulence structures (Mauder et al. 2020). This does not mean that the Bowen ratio method is more accurate, because it distributes the proportion of energy flux through large turbulence structures between the sensible and latent heat fluxes according to the Bowen ratio, which is not necessarily correct, because the sensible heat flux is often more strongly influenced.

The ***Bowen ratio method*** is one of the most common methods used to determine the fluxes of sensible and latent heat. For this purpose, the net radiation (see ▶ Sect. 3.4.3) and the ground heat flow (see ▶ Sect. 3.8) must be measured. The Bowen ratio is determined from the temperature difference and the vapour pressure difference at two heights. Psychrometers (see ▶ Sect. 3.2.3) or ventilated thermometers (see ▶ Sect. 3.1) and dew point hygrometers (see ▶ Sect. 3.2.3) should be used for this purpose. An example of a device is shown in Fig. 3.24. The measuring heights should be determined in such a way that the ratio of both heights measured above the displacement height is at least 4–8 to achieve the highest possible accuracy. Furthermore, at least one wind measurement (see ▶ Sect. 3.3.3) should be carried out, as the method can only be used if turbulence has developed. The evaluation algorithm is shown in Note 3.10.

Note 3.10

Calculation of the Bowen ratio method: The method is based on the definition of the Bowen ratio

$$Bo = \frac{Q_H}{Q_E} \tag{N3.12}$$

with the sensible heat flux Q_H and the latent het flux Q_E and the energy balance equation

$$-Q_s^* = Q_H + Q_E + Q_G \tag{N3.13}$$

with the net radiation $-Q_s^*$ and the ground heat flux Q_G. Furthermore, the Bowen ratio similarity is also applied (Foken and Mauder 2024)

$$Bo = \gamma \frac{\Delta T}{\Delta e} \tag{N3.14}$$

where $\gamma = 0.667$ hPa K^{-1} for $p = 1013.25$ hPa and $t = 20$ °C is the psychrometric constant. For other pressure and temperature values, the psychrometer constant is tabulated in Sonntag et al. (2021). Furthermore, ΔT and Δe are the temperature and vapour pressure differences between two heights. From both equations, the sensible and latent heat fluxes can be determined:

$$Q_H = (-Q_s^* - Q_G)\frac{Bo}{1 + Bo} \tag{N3.15}$$

$$Q_E = \frac{-Q_s^* - Q_G}{1 + Bo}. \tag{N3.16}$$

The equation cannot be solved for Bo = –1 and the method is therefore not applicable for sensible and latent heat fluxes of the same magnitude, but with different signs, and concerns the morning and evening hours of about 1–2 h duration.

Special estimates are necessary to determine the sign of the fluxes at Bo < 0 (Ohmura 1982):

$$\text{If } \left(-Q_s^* - Q_G\right) > 0 \text{ then} \left(\lambda \Delta q + c_p \Delta T\right) > 0 \tag{N3.17}$$

$$\text{If } \left(-Q_s^* - Q_G\right) < 0 \text{ then} \left(\lambda \Delta q + c_p \Delta T\right) < 0 \tag{N3.18}$$

where λ is the heat of vaporisation, q the specific humidity, and c_p is the specific heat capacity of moist air at constant pressure.

To ensure that a sufficient turbulent regime exists, it is recommended that only measurements with a wind velocity at the upper height greater than 1 m s^{-1} and/or a difference of the wind velocities between both heights greater than 0.3 m s^{-1} should be used. Unrealistic energy fluxes are predicted for the morning and evening hours. Therefore, the range $-1.25 < Bo < -0.75$ should be excluded from further analysis. Error analyses for the Bowen ratio method are available; see Foken and Mauder (2024).

In the ***eddy-covariance method***, the fluxes are determined from the covariance of the turbulent fluctuations of the vertical wind velocity and horizontal wind velocity (momentum flux), air temperature (sensible heat flux), humidity (latent heat flux), or concentrations of trace gases (e.g. carbon dioxide flux). The input variables must be measured with a very high temporal resolution of 10–20 Hz, whereby the measuring devices must be nearly inertia-free. Sonic anemometers (see ► Sect. 3.3.3) and optical measuring devices for water vapour (see ► Sect. 3.2.3) and other gases are therefore used. The fluxes are then mainly determined for 30-min averages. Since these devices have been available with the necessary temporal resolution since the 1960s, the eddy-covariance method has developed into one of the most widely used micrometeorological measurement methods.

The application of the method requires a number of prerequisites that must be checked during a quality analysis. These include the fact that the mean vertical wind in the measurement interval must be zero, which can be realized by coordinate rotation. Furthermore, a number of corrections must be made, such as the density influence due to the water vapour content and the failure to determine high-frequency fluctuations. See the relevant technical literature for these corrections and others not mentioned here (Aubinet et al. 2012; Mauder et al. 2021). Although commercial analysing programmes are available today, the application requires a great level of detailed knowledge, so you should make use of the training courses on offer. Burba (2022) gave a simple introduction that is available online.

The ***modified Bowen ratio method*** generally assumed that the ratio of two turbulent fluxes is the same as the ratio of their vertical gradients (Note 3.11). It is therefore also possible to directly determine a flux that is easy to measure using the eddy-covariance method, for example, and to determine the

3

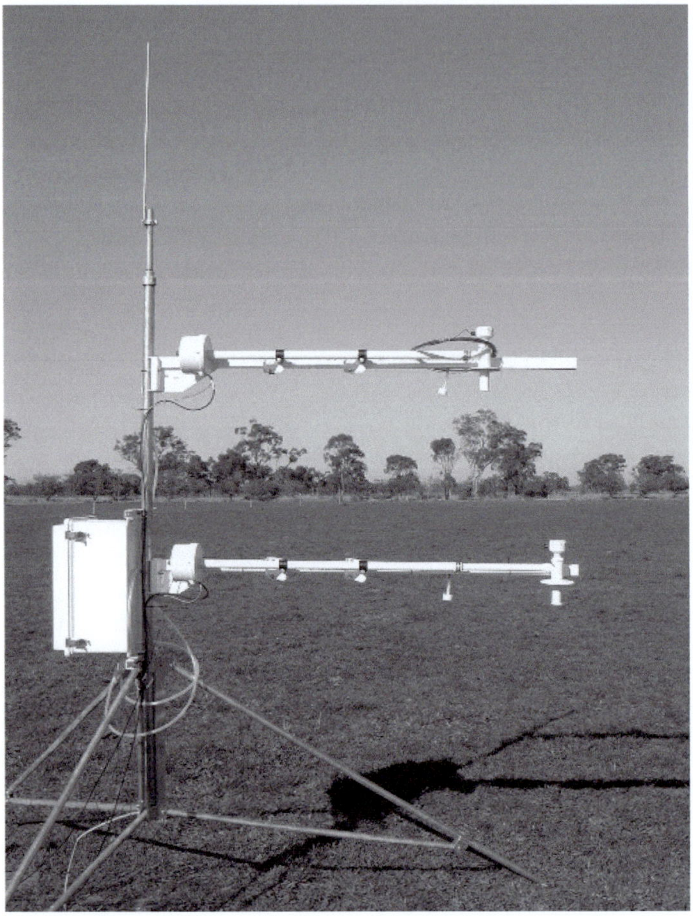

Fig. 3.24 Bowen ratio measurement system (Published with kind permission of © Campbell Scientific, Inc. Logan UT, USA, All rights reserved)

flux that is difficult to determine using Eq. (N3.17). If, for example, the sensible heat flux is determined using the eddy-covariance method and the temperature and vapour pressure differences between two heights, the latent heat flux can be calculated. In contrast to the Bowen ratio method, it is not necessary to measure the net radiation and the ground heat flux (Liu and Foken 2001). It is far more common to measure the difference of a trace gas concentration, so that the flux of the trace gas can then be determined using this method (Businger 1986). The prerequisite is that both fluxes behave similarly, i.e. have a similar diurnal cycle, for example.

Note 3.11

Modified Bowen ratio method:

Analogous to the Bowen ratio similarity (Eqs. N3.12 and N3.14), the following relationship can also be determined:

$$\frac{Q_a}{Q_b} = \frac{\Delta a}{\Delta b}. \quad \text{(N3.17)}$$

Q_a and Q_b are the fluxes of the quantities a and b.

Scintillometer measurements are a relatively new technique in micrometeorology. Following the development of wave propagation in

Table 3.30 Typical parameters of scintillometers

Type	Short name	Wavelength	Path length	Path height
(displaced beam) small-aperture scintillometer	(DB)SAS	0.7–0.8 µm	0.05–0.25 km	0.5–5 m
(extra) large-aperture scintillometer	(X)LAS	0.8–0.9 µm	0.25–4.5 km XLAS 10 km	1.5–50 (100) m
Microwave scintillometer		1.8–1.9 mm	1–10 km	3–100 m

the atmosphere in the middle of the last century, the first devices were developed in the 1970s (Hill et al. 1980). These utilize the scintillation of light due to density inhomogeneities in the air as a result of temperature or humidity differences. The measurement is carried out along a measuring path at a specific height depending on the device type. A distinction is made between the three types of scintillometers shown in Table 3.30 (Ward 2017; Beyrich et al. 2021).

Using complex models based on similarity theories of atmospheric turbulence, the sensible heat flux can be determined with SAS and LAS, the friction velocity with SAS, and the latent heat flux with microwave scintillometers. Scintillometers are not able to determine the sign of the sensible heat flux. Additional measurements (temperature gradient; see ▶ Sect. 3.12.3) are necessary, if the sign cannot simply be determined by the daily cycle because possibly unclear situations can occur during the afternoon transition (see ▶ Sect. 1.2.7). Scintillometers have a sensitivity that is in the middle of the measurement path rather than near the transmitter and receiver. This must be taken into account for footprint analyses of the measurement sector (Meijninger et al. 2002). SAS are suitable for short-term measurement programmes; the other two types are more suitable for longer term stationary programmes. In any case, an expert should be consulted for installation and interpretation.

3.13 Chamber Methods

Ecologists use soil chambers to investigate the matter exchange of gases between the atmosphere and the soil. With this method fluxes can be determined for very small plots. One of the oldest chamber types and nearly 100 years in use is the non-flow-through steady-state chamber. These chambers are fully closed with an absorber inside. As an example, the respired carbon dioxide can be measured with an alkali absorber. After some hours or days, the absorber must be chemically analysed. A good overview about this technique is given by Rochette and Hutchinson (2005) as well as Subke et al. (2021). In addition to chamber measurements on the soil, chambers can also be attached to plant parts that work in a similar way (Perez-Priego 2021).

3.13.1 Why Should Chamber Measurements Be Used?

Chamber measurements are always used where the measuring areas are very small and other methods cannot be used due to the large footprint area. In addition, a large number of gases can be measured better with chambers, as no analysers are available, particularly with a high temporal resolution, or low concentrations can only be measured inaccurately. Table 3.31 shows typical application examples.

3.13.2 Where to Measure

In principle, there are no restrictions for applications a–c, measuring over-vegetated (very low vegetation) or unvegetated soil. It is important that the chamber is placed close to the soil. For this purpose, rings are usually inserted into the soil, on which the chamber is placed airtight. This is particularly im-

Table 3.31 Application purpose for chamber measurements

Code	Application	Relevance
a	Flux measurements on small plots	
b	Comparison of fluxes on different plots	
c	Flux measurements of gases that are difficult to measure with other methods	

portant for manual measurements where the chamber is inserted at different points. This sporadic chamber measurements are based on the assumption of a constant emission flux from the soil into the atmosphere. Thereby it is ignored that the pressure field, the stratification of the atmosphere, and the turbulence can modify the emission rate, at least the emission at the top of the surface. For this reason, non-continuous measurements should be carried out at short intervals, if possible, for example in order to obtain a good picture of the diurnal cycle.

3.13.3 Which Chamber Method is to Be Used for Measurements

Chambers separate the near-surface volume from the atmosphere such that typical atmospheric turbulent conditions are not present within the chamber and the longwave net radiation inside the chamber is nearly zero. To overcome these problems, the chamber is often ventilated, but the typical wind profile of a boundary layer cannot be realized. Cooling pads are used to prevent overheating due to shortwave radiation. Furthermore, non-transparent chambers can measure the respiration at daytime too (Table 3.32). To reduce errors under atmospheric conditions, parallel measurements are recommended with eddy-covariance techniques on a nearby larger plot with a chamber in the maximum of the footprint for comparison.

Obviously, in the range of a good agreement with eddy-covariance measurements the ventilation of the chamber is similar to the conditions outside the chamber. The adjusted longwave net radiation inside the chamber is often very much underestimated, while outside large values are possible with cooling at the surface resulting in stable stratification. Under these conditions, during the night or during the oasis effect in the late afternoon (see ▶ Sect. 1.2.7), chambers overestimate the fluxes significantly. But in the case of coherent structures (see ▶ Sect. 2.2), which realize a well-mixed situation over a short period, even at night, smaller fluxes are measured in the chamber (Riederer et al. 2014; Brændholt et al. 2017).

The non-flow-through steady-state chamber is however nowadays rarely used except for isotope techniques. Non-steady-state chambers are now the most common. Thereby, in the case of emission from the soil, the concentration inside the chamber increases in a time interval of some minutes, while the concentration is continuously measured. From the slope of the concentration increase, the emission can be calculated. To realize a linear slope, the chamber must be ventilated or equipped with a pressure vent at the top of the chamber (Xu et al. 2006). Another system, the flow-through non-steady-state chamber, works with a constant airflow through the ventilated chamber, where the concentration is measured at the in- and outflow. The benefit is a high time resolution of the measurements, but the gas analysers need a high accuracy and sampling frequency. Accordingly, these chambers are only available for some gases. A novel *steady-state* approach measures the rate of CO_2 diffusion across a permeable membrane to establish soil CO_2 efflux based on the CO_2 concentration within an otherwise closed chamber headspace (forced diffusion chamber, Table 3.32).

There are a large number of chambers that have been produced in research institutes. Before using them, you should familiarize yourself with the results of chamber comparisons (Pumpanen et al. 2004; Pihlatie et al. 2013). In most cases, instruction by a specialist is essential. Commercial chambers are easy to use but are relatively expensive and are available for both manual (Fig. 3.25) and automatic operation.

Table 3.32 Advantages and limitations of the different chamber configurations (Perez-Priego 2021; Subke et al. 2021)

Operating conditions	Advantages	Limitations
Closed dynamic	Simple, portable, and suitable for field conditions, adaptable (transparent and opaque for photosynthesis and respiration) Instantaneous during an optimal period of enclosure at low ventilation rates can be achieved	An operator is required but unattended systems are also possible Disturbances and artefacts during the enclosure periods (e.g. physiology, non-steady-state conditions with increasing humidification and saturation problems, dew formation, leakages)
Open dynamic	Continuous and repeated measurements, allows flux measurements under controlled conditions, capacity to hold unvarying environmental conditions during the enclosure period	Requires heavy equipment, Ventilation regimes can be highly altered by gas chromatograph Complex operation (particularly under intermittent environmental conditions), pressure compensation
Static	Simple to use, adaptive (transparent and opaque for photosynthesis and respiration) Easily collect the air sampling into an evacuate tube	Underestimate the fluxes, Laboratory background (gas chromatograph, etc.), not continuous measurement
Manual	Few spatial constraints due to low weight and small size Versatile for comparative analyses across ecosystems and treatments	Manual operation, limited by battery power, limited areal, and temporal replication
Automated	Low demand on person time, high temporal resolution, can be operated in response to specific environmental triggers/events	Significant maintenance demand for continuous operation, risk of mechanical failure, limited in spatial reach, requires secure power supply
Forced diffusion	No spatial limitation Continuous data collection Very low power demand Reduced maintenance requirements due the absence of moving parts	Potential impacts on abiotic and biotic conditions in the case of prolonged placement of the same location, separation of the soil from the surrounding

Fig. 3.25 Commercially available manual chamber © LI-COR

3

References

Allen RG, Pereira LS, Raes D, Smith M (1998) Crop evaporation. FAO Irrigation Drainage Pap 56:XXVI + 300 pp.

Allen R, Foken T, Kilic A, Trezza R, Ortega Farias AS (2021a) Evapotranspiration measurements and calculations. In: Foken T (ed) Springer handbook of atmospheric measurements. Springer, Cham, pp 1531–1567. ▸ https://doi.org/10.1007/978-3-030-52171-4_57

Allen RG, Dhungel R, Dhungana B, Huntington J, Kilic A, Morton C (2021b) Conditioning point and gridded weather data under aridity conditions for calculation of reference evapotranspiration. Agric Water Manag 245:106531. ▸ https://doi.org/10.1016/j.agwat.2020.106531

Aubinet M, Vesala T, Papale D (Hrsg) (2012) Eddy covariance: a practical guide to measurement and data analysis. Springer, Dordrecht, Heidelberg, London, New York. ▸ https://doi.org/10.1007/978-94-007-2351-1

Baldocchi D, Falge E, Gu L, Olson R, Hollinger D, Running S, Anthoni P, Bernhofer C, Davis K, Evans R, Fuentes J, Goldstein A, Katul G, Law B, Lee XH, Malhi Y, Meyers T, Munger W, Oechel W, Paw UKT, Pilegaard K, Schmid HP, Valentini R, Verma S, Vesala T (2001) FLUXNET: a new tool to study the temporal and spatial variability of ecosystem-scale carbon dioxide, water vapor, and energy flux densities. Bull Amer Meteorol Soc 82:2415–2434. ▸ https://doi.org/10.1175/1520-0477(2001)082%3c2415:FANTTS%3e2.3.CO;2

Behrens K (2021) Radiation sensors. In: Foken T (ed) Springer Handbook of atmospheric measurements. Springer, Cham, pp 279–357. ▸ https://doi.org/10.1007/978-3-030-52171-4_11

Beyrich F, Hartogensis OK, de Bruin HAR, Ward HC (2021) Scintillometers. In: Foken T (ed) Springer handbook of atmospheric measurements. Springer International Publishing, Cham, pp 969–997. ▸ https://doi.org/10.1007/978-3-030-52171-4_34

Bonan G (2019) Soil temperature. In: Bonan G (Hrsg) Climate change and terrestrial ecosystem modeling. Cambridge University Press, Cambridge, pp 64–79. ▸ https://doi.org/10.1017/9781107339217.006

Brændholt A, Steenberg Larsen K, Ibrom A, Pilegaard K (2017) Overestimation of closed-chamber soil CO_2 effluxes at low atmospheric turbulence. Biogeoscience 14(6):1603–1616. ▸ https://doi.org/10.5194/bg-14-1603-2017

Brock FV, Richardson SJ (2001) Meteorological measurement systems. Oxford University Press, New York. ▸ https://doi.org/10.1093/oso/9780195134513.001.0001

Bröde P, Blazejczyk K, Fiala D, Havenith G, Holmér I, Jendritzky G, Kuklane K, Kampmann B (2013) The universal thermal climate index UTCI compared to ergonomics standards for assessing the thermal environment. Ind Health 51(1):16–24. ▸ https://doi.org/10.2486/indhealth.2012-0098

Brotzge JA, Duchon CE (2000) A field comparison among a Domeless net radiometer, two four-component net radiometers, and a domed net radiometer. J Atm Oceanic Technol 17(12):1569–1582. ▸ https://doi.org/10.1175/1520-0426(2000)017%3c1569:AFCAAD%3e2.0.CO;2

Burba G (2022) Eddy covariance method for scientific, regulatory, and commercial applications. Lincoln, LI-COR Biosciences

Businger JA (1986) Evaluation of the accuracy with which dry deposition can be measured with current micrometeorological techniques. J Appl Meteorol 25:1100–1124. ▸ https://doi.org/10.1175/1520-0450(1986)025%3c1100:EOTAWW%3e2.0.CO;2

Cauteruccio A, Colli M, Stagnaro M, Lanza LG, Vuerich E (2021) In situ precipitation measurements. In: Foken T (Ed) Springer handbook of atmospheric measurements. Springer Nature, Cham, pp 359–400. ▸ https://doi.org/10.1007/978-3-030-52171-4_12

de Vries DA (1963) Thermal properties of soils. In: van Wijk WR (Hrsg) Physics of the plant environment. North-Holland Publ. Co., Amsterdam, pp 210–235

DeBruin HAR, Holtslag AAM (1982) A simple parametrization of the surface fluxes of sensible and latent heat during daytime compared with the Penman-Monteith concept. J Climate Appl Meteorol 21:1610–1621. ▸ https://doi.org/10.1175/1520-0450(1982)021%3c1610:ASPOTS%3e2.0.CO;2

DeFelice TP (1998) An introduction to meteorological instrumentation and measurement. Prentice Hall, Upper Saddle River

Dey S, Amin EM, Karmakar NC (2020) Paper based chipless RFID leaf wetness detector for plant health monitoring. IEEE Access 8:191986–191996. ▸ https://doi.org/10.1109/ACCESS.2020.3033191

DIN-EN-IEC (2023) Industrielle Platin-Widerstandsthermometer und platin-temperatursensoren (Industrial platinum resistance thermometers and platinum temperature sensors, IEC 60751:2022). Beuth Verlag, Berlin. ▸ https://doi.org/10.31030/3405985

DIN-EN (2009) Temperatur - Tabellen der elektromotorischen Kraft (EMK) für Kombinationen von Reinelement-Thermoelementen (Temperature—Electromotive force (EMF) tables for pure-element thermocouple combinations, IEC 62460:2008), DIN EN 62460. Beuth Verlag, Berlin. ▸ https://doi.org/10.31030/1507752

DIN-ISO (2003) Meteorologie - Ultraschall-Anemometer/Thermometer - Abnahmeprüfverfahren für Messungen der mittleren Windgeschwindigkeit (Meteorology-Sonic anemometer/thermometer—Acceptance test method for mean wind measurements, ISO 16622). Beuth-Verlag, Berlin. ▸ https://doi.org/10.31030/9429300

DIN-ISO (2007) Meteorologie - Windmessungen - Teil 1: Prüfverfahren in Windkanälen zur Ermittlung der Leistung von Rotationsanemometern (Meteorology—Wind measurements—Part 1: wind tunnel test methods for rotating anemometer performance, ISO 17713-1), ISO 17713-1. Beuth-Verlag, Berlin. ▸ https://doi.org/10.31030/9866977

Doorenbos J, Pruitt WO (1977) Guidelines for predicting crop water requirements, 2nd edn. FAO Irrigation Drainage Pap 24, 145 pp.

Emeis S (2010) Measurement methods in atmospheric sciences. Borntraeger Science Publishers, Stuttgart

Emeis S, Wilbert S (2021) Measurement systems of wind, solar and hydropower applications. In: Foken T (ed) Springer handbook of atmospheric measurements. Springer, Cham, pp 1371–1391. ► https://doi.org/10.1007/978-3-030-52171-4_51

Farahani H, Wagiran R, Hamidon MN (2014) Humidity sensors principle, mechanism, and fabrication technologies: a comprehensive review. Sensors 14:7881–7939. ► https://doi.org/10.3390/s140507881

Foken T (ed) (2021) Springer handbook of atmospheric measurements. Springer, Cham. ► https://doi.org/10.1007/978-3-030-52171-4

Foken T, Bange J (2021a) Temperature sensors. In: Foken T (ed) Springer handbook of atmospheric measurements. Springer, Cham, pp 183–208. ► https://doi.org/10.1007/978-3-030-52171-4_7

Foken T, Bange J (2021b) Wind sensors. In: Foken T (ed) Springer handbook of atmospheric measurements. Springer, Cham, pp 243–272. ► https://doi.org/10.1007/978-3-030-52171-4_9

Foken T, Hellmuth O, Huwe B, Sonntag D (2021) Physical quantities. In: Foken T (ed) Springer handbook of atmospheric measurements. Springer Nature, Cham, pp 107–151. ► https://doi.org/10.1007/978-3-030-52171-4_5

Foken T, Mauder M (2024) Micrometeorology, 3edn. Springer, Cham. ► https://doi.org/10.1007/978-3-031-47526-9

Gao Z, Russell Eric S, Missik Justine EC, Huang M, Chen X, Strickland Chris E, Clayton R, Arntzen E, Ma Y, Liu H (2017) A novel approach to evaluate soil heat flux calculation: an analytical review of nine methods. J Geophys Res: Atmosph 122(13):6934–6949. ► https://doi.org/10.1002/2017JD027160

Gilgen H, Whitlock CH, Koch F, Müller G, Ohmura A, Steiger D, Wheeler R (1994) Technical plan for BSRN data management. WRMC, Technol Rep 1:56 pp.

Hanafusa T, Fujitana T, Kobori Y, Mitsuta Y (1982) A new type sonic anemometer-thermometer for field operation. Papers Meteorol Geophys 33:1–19. ► https://doi.org/10.2467/mripapers.33.1

Harrison GR (2015) Meteorological measurements and instrumentation. Wiley, Chichester. ► https://doi.org/10.1002/9781118745793

Hebra AJ (2010) The physics of metrology. Springer, Wien, New York. ► https://doi.org/10.1007/978-3-211-78381-8

Hill RJ, Clifford SF, Lawrence RS (1980) Refractive index and absorption fluctuations in the infrared caused by temperature, humidity and pressure fluctuations. J Opt Soc Am 70:1192–1205. ► https://doi.org/10.1364/JOSA.70.001192

Huwe B, Bogner C, Foken T (2021) Soil measurements. In: Foken T (ed) Springer handbook of atmospheric measurements. Springer, Cham, pp 1627–1658. ► https://doi.org/10.1007/978-3-030-52171-4_61

IEC (2017) Wind energy generation systems—Part 12-1: power performance measurements of electricity producing wind turbines, IEC 61400-12-1, Edition 2.0, 2017-03. VDE Verlag, Berlin

ISO (2007) Optics and photonics—Spectral bands, ISO 20473. ISO copyright office, Geneva

ISO (2015) Air quality—Meteorology—Siting classifications for surface observing stations on land, ISO 19289:2015 (WMO)

ISO (2018) Solar energy—Specification and classification of instruments for measuring hemispherical solar and direct solar radiation, ISO 9060. ISO, Geneva

Jakobi J, Huisman JA, Köhli M, Rasche D, Vereecken H, Bogena HR (2021) The footprint characteristics of cosmic ray thermal neutrons. Geophys Res Letters 48(15):e2021GL094281. ► https://doi.org/10.1029/2021GL094281

Kristensen L (1998) Cup anemometer behavior in turbulent environments. J Atm Oceanic Techn 15:5–17. ► https://doi.org/10.1175/1520-0426(1998)015%3c0005:CABITE%3e2.0.CO;2

Lalic B, Eitzinger J, Marta AD, Sremac AF, Orlandini S, Pacher B (2018) Agricultural meteorology and climatology. Firenze University Press, Firenze

Latimer JR (1972) Radiation measurement, International Field Year of the Great Lakes, Techn. Manual Series No. 2, Information. In. Ottawa, p 53

Leclerc MY, Foken T (2014) Footprints in micrometeorology and ecology. Springer, Heidelberg, New York, Dordrecht, London. ► https://doi.org/10.1007/978-3-642-54545-0

Liebethal C, Huwe B, Foken T (2005) Sensitivity analysis for two ground heat flux calculation approaches. Agric Meteorol 132:253–262. ► https://doi.org/10.1016/j.agrformet.2005.08.001

Liebethal C, Foken T (2006) On the use of two repeatedly heated sensors in the determination of physical soil parameters. Meteorol Z 15:293–299. ► https://doi.org/10.1127/0941-2948/2006/0125

Liebethal C, Foken T (2007) Evaluation of six parameterization approaches for the ground heat flux. Theor Appl Climat 88:43–56. ► https://doi.org/10.1007/s00704-005-0234-0

Liu H, Foken T (2001) A modified Bowen ratio method to determine sensible and latent heat fluxes. Meteorol Z 10:71–80. ► https://doi.org/10.1127/0941-2948/2001/0010-0071

Lu Y-C, Romps DM (2022) Extending the heat index. J Appl Meteorol Climatol 61(10):1367–1383. ► https://doi.org/10.1175/JAMC-D-22-0021.1

Mauder M, Genzel S, Fu J, Kiese R, Soltani M, Steinbrecher R, Zeeman M, Banerjee T, De Roo F, Kunstmann H (2018) Evaluation of energy balance closure adjustment methods by independent evapotranspiration estimates from lysimeters and hydrological simulations. Hydrol Processes 32(1):39–50. ► https://doi.org/10.1002/hyp.11397

3

Mauder M, Foken T, Cuxart J (2020) Surface energy balance closure over land: a review. Boundary-Layer Meteorol 177:395–426. ▶ https://doi.org/10.1007/s10546-020-00529-6'

Mauder M, Foken T, Aubinet M, Ibrom A (2021) Eddy-covariance measurements. In: Foken T (ed) Springer handbook of atmospheric measurements. Springer Nature, Cham, pp 1473–1504. ▶ https://doi.org/10.1007/978-3-030-52171-4_55

Meijninger WML, Green AE, Hartogensis OK, Kohsiek W, Hoedjes JCB, Zuurbier RM, DeBruin HAR (2002) Determination of area-averaged water vapour fluxes with large aperture and radio wave scintillometers over a heterogeneous surface—Flevoland field experiment. Boundary-Layer Meteorol 105:63–83. ▶ https://doi.org/10.1023/A:1019683616097

Moene AF, van Dam JC (2014) Transport in the atmosphere-vegetation-soil continuum. Cambridge University Press, Cambridge. ▶ https://doi.org/10.1017/CBO9781139043137

Monteith JL (1965) Evaporation and environment. Symp Soc Exp Biol 19:205–234

Nguyen BH, Gilbert GS, Rolandi M (2023) A bio-mimetic leaf wetness sensor from replica molding of leaves. Adv Sens Res 2(6):2200033. ▶ https://doi.org/10.1002/adsr.202200033

Ohmura A (1982) Objective criteria for rejecting data for Bowen ratio flux calculations. J Climate Appl Meteorol 21:595–598. ▶ https://doi.org/10.1175/1520-0450(1982)021%3c0595:OCFRDF%3e2.0.CO;2

Op de Beeck M, Sabbatini S, Papale D (2017) ICOS ecosystem instructions for soil meteorological measurements (TS, SWC, G) (Version 20180615). ICOS Ecosystem Thematic Centre, Viterbo. ▶ https://doi.org/10.18160/1a28-gex6

Penman HL (1948) Natural evaporation from open water, bare soil and grass. Proc R Soc London A193:120–195. ▶ https://doi.org/10.1098/rspa.1948.0037

Perez-Priego O (2021) Plant chamber measurements. In: Foken T (ed) Springer handbook of atmospheric measurements. Springer International Publishing, Cham, pp 1585–1601. ▶ https://doi.org/10.1007/978-3-030-52171-4_59

Philip JR (1961) The theory of heat flux meters. J Geophys Res 66:571–579. ▶ https://doi.org/10.1029/JZ066i002p00571

Piche A (1872) Note sur l'atmismomètre instrument destiné a mesurer l'évaporation. Bulletin Association Scientifique De France 10:166–168

Pihlatie MK, Christiansen JR, Aaltonen H, Korhonen JFJ, Nordbo A, Rasilo T, Benanti G, Giebels M, Helmy M, Sheehy J, Jones S, Juszczak R, Klefoth R, Lobo-do-Vale R, Rosa AP, Schreiber P, Serça D, Vicca S, Wolf B, Pumpanen J (2013) Comparison of static chambers to measure CH_4 emissions from soils. Agric For Meteorol 171–172(0):124–136. ▶ https://doi.org/10.1016/j.agrformet.2012.11.008

Pirasteh-Anosheh H, Saed-Moucheshi A, Pakniyat H, Pessarakli M (2016) Stomatal responses to drought stress. In: Ahmad P (Hrsg) Water stress and crop plants, pp 24–40. ▶ https://doi.org/10.1002/9781119054450.ch3

Preston-Thomas H (1990) The international temperature scale of 1990 (ITS-90). Metrologia 27:3–10. ▶ https://doi.org/10.1088/0026-1394/27/1/002

Priestley CHB, Taylor JR (1972) On the assessment of surface heat flux and evaporation using large-scale parameters. Mon Weather Rev 100:81–92. ▶ https://doi.org/10.1175/1520-0493(1972)100%3c0081:OTAOSH%3e2.3.CO;2

Pumpanen J, Kolari P, Ilvesniemi H, Minkkinen K, Vesala T, Niinistö S, Lohila A, Larmola T, Morero M, Pihlatie M, Janssens I, Yuste JC, Grünzweig JM, Reth S, Subke J-A, Savage K, Kutsch W, Østreng G, Ziegler W, Anthoni P, Lindroth A, Hari P (2004) Comparison of different chamber techniques for measuring soil CO_2 efflux. Agric Meteorol 123(3–4):159–176. ▶ https://doi.org/10.1016/j.agrformet.2003.12.001

Reth S, Perez-Priego O, Coners H, Nolz R (2021) Lysimeter. In: Foken T (ed) Springer handbook of atmospheric measurements. Springer International Publishing, Cham, pp 1569–1584. ▶ https://doi.org/10.1007/978-3-030-52171-4_58

Riederer M, Serafimovich A, Foken T (2014) Eddy covariance—Chamber flux differences and its dependence on atmospheric conditions. Atmos Meas Technol 7:1057–1064. ▶ https://doi.org/10.5194/amt-7-1057-2014

Robinson DA, Jones SB, Wraith JM, Or D, Friedman SP (2003) A Review of advances in dielectric and electrical conductivity measurement in soils using time domain reflectometry. Vadose Zone J 2(4):444–475. ▶ https://doi.org/10.2136/vzj2003.4440

Robinson DA, Campbell CS, Hopmans JW, Hornbuckle BK, Jones SB, Knight R, Ogden F, Selker J, Wendroth O (2008) Soil moisture measurement for ecological and hydrological watershed-scale observatories: a review. Vadose Zone J 7(1):358–389. ▶ https://doi.org/10.2136/vzj2007.0143

Rochette P, Hutchinson GL (2005) Measurement ofsoil respiration in situ: chamber techniques. In: Hatfield JL, Baker I (Hrsg) Micrometeorology in agricultural systems, Bd 47. American Society of Agronomy, Madison, pp 247–286

Rowlandson T, Gleason M, Sentelhas P, Gillespie T, Thomas C, Hornbuckle B (2014) Reconsidering leaf wetness duration determination for plant disease management. Plant Dis 99(3):310–319. ▶ https://doi.org/10.1094/PDIS-05-14-0529-FE

Sauer TJ, Horton R (2005) Soil heat flux. In: Hatfield JL, Baker JM (Hrsg) Micrometeorology in agricultural systems. American Society of Agronomy, Inc., Madison, pp 131–134

Schmid HP, Rebmann C (2021) Integration of meteorological and ecological measurements. In: Foken T (ed) Springer handbook of atmospheric measurements. Springer International Publishing, Cham, pp 1713–1725. ▶ https://doi.org/10.1007/978-3-030-52171-4_64

Schmitz HF, Grant RH (2009) Precipitation and dew in a soybean canopy: spatial variations in leaf wetness and implications for Phakopsora pachyrhizi infection. Agric For Meteorol 149(10):1621–1627. ► https://doi.org/10.1016/j.agrformet.2009.05.001

Seltmann JEE (2021) Weather radar. In: Foken T (ed) Springer handbook of atmospheric measurements. Springer International Publishing, Cham, pp 841–900. ► https://doi.org/10.1007/978-3-030-52171-4_30

Shuttleworth WJ, Wallace JS (1985) Evaporation from sparse crops-an energy combination theory. Quart J Roy Meteorol Soc 111(469):839–855. ► https://doi.org/10.1002/qj.49711146910

Smajstrla AG, Zazueta FS, Clark GA, Pitts DJ (2000) Irrigation scheduling with evaporation pans. Univ of Florida, IFAS Ext Bul 254:9

Sonntag D (1989) WMO Assmann aspiration psychrometer intercomparison. WMO, Instruments Observ Methods 34:185

Sonntag D (1990) Important new values of the physical constants of 1986, vapour pressure formulations based on the ITS-90, and psychrometer formulae. Z Meteorol 40:340–344

Sonntag D (1994) Advancements in the field of hygrometry. Meteorol Z 3:51–66. ► https://doi.org/10.1127/metz/3/1994/51

Sonntag D, Foken T, Vömel H, Hellmuth O (2021) Humidity sensors. In: Foken T (ed) Springer handbook of atmospheric measurements. Springer, Cham, pp 209–242. ► https://doi.org/10.1007/978-3-030-52171-4_8

Stull RB (1988) An introduction to boundary layer meteorology. Kluwer, Dordrecht. ► https://doi.org/10.1007/978-94-009-3027-8

Subke J-A, Kutzbach L, Risk D (2021) Soil chamber measurements. In: Foken T (ed) Springer handbook of atmospheric measurements. Springer International Publishing, Cham, pp 1603–1624. ► https://doi.org/10.1007/978-3-030-52171-4_60

Szilagyi J, Parlange MB, Katul G (2015) Assessment of the Priestley-Taylor parameter value from ERA-interim global reanalysis data. J Hydrol Environ Res 2:1–7

Szilagyi J, Crago RD (2023) A thermodynamics-based versatile evapotranspiration estimation method of minimum data requirement for water resources investigations. J Hydrol 624:129917. ► https://doi.org/10.1016/j.jhydrol.2023.129917

Torri A, Bange J, Foken T (2021) Pressure sensors. In: Foken T (ed) Springer handbook of atmospheric measurements. Springer, Cham, pp 273–295. ► https://doi.org/10.1007/978-3-030-52171-4_10

Trebs I, Ammann C, Junk J (2021) Immission and dry deposition. In: Foken T (ed) Springer handbook of atmospheric measurements. Springer International Publishing, Cham, pp 1445–1471. ► https://doi.org/10.1007/978-3-030-52171-4_54

Turc L (1961) Évaluation des besoins en eau d'irrigation évapotranspiration potentielle. Ann Agron 12:13–49

VDI (2018) Umweltmeteorologie, Meteorologische Messungen, Wind (Environmental meteorology, meteorological measurements, wind), VDI 3786, Blatt (Part) 2. Beuth Verlag, Berlin

VDI (2024) Umweltmeteorologie, Meteorologische Messungen, Lufttemperatur (Environmental meteorology, Meteorological Measurements, Air temperature), VDI 3786 Blatt 3 (Part 3). Beuth Verlag, Berlin

Vulova S, Rocha AD, Meier F, Nouri H, Schulz C, Soulsby C, Tetzlaff D, Kleinschmit B (2023) City-wide, high-resolution mapping of evapotranspiration to guide climate-resilient planning. Rem Sens Environm 287:113487. ► https://doi.org/10.1016/j.rse.2023.113487

Ward HC (2017) Scintillometry in urban and complex environments: a review. Meas Sci Technol 28(6):064005. ► https://doi.org/10.1088/1361-6501/aa5e85

Wendisch M, Yang P (2012) Theory of atmospheric radiative transfer. Wiley, Weinheim

Wessel-Bothe S, Weihermüller L (eds) (2020) Field measurement methods in soil science. Borntraeger Science Publisher, Stuttgart

Wichura B, Foken T (2021) Atmospheric measurements for different purposes. In: Foken T (ed) Springer handbook of atmospheric measurements. Springer, Cham, pp 1187–1197. ► https://doi.org/10.1007/978-3-030-52171-4_43

WMO (2020) Guide to hydrological practice, Volume I, Hydrology—From measurement to hydrological information, WMO No. 168. World Meteorological Organization, Geneva

WMO (2024) Guide to instruments and methods of observation, WMO-No. 8, Volume I—Measurement of Meteorological Variables. World Meteorological Organization, Geneva

Xu L, Furtaw MD, Madsen RA, Garcia RL, Anderson DJ, McDermitt DK (2006) On maintaining pressure equilibrium between a soil CO_2 flux chamber and the ambient air. J Geophys Res: Atmosph 111(D8):D08S10. ► https://doi.org/10.1029/2005JD006435

Yang K, Wang J (2008) A temperature prediction-correction method for estimating surface soil heat flux from soil temperature and moisture data. Sci China Ser D: Earth Sci 51:721–729. ► https://doi.org/10.1007/s11430-008-0036-1

Quality Control and Recovery of Meteorological Data

Mina Petrić, Branislava Lalic, Peter Domonkos, Tamás Weidinger, Thomas Vergauwen, Ivan Koči and Thomas Foken

B. Lalic et al. (eds.), *Micrometeorological Measurements*,
Springer Textbooks in Earth Sciences, Geography and Environment,
https://doi.org/10.1007/978-3-032-03884-5_4

4

Meteorological measurements require significant time, human effort, and resources, including instrumental and operational costs. The resulting meteorological database must have a reliable consistency, which must be designed according to the user's needs. The whole measurement process, from planning and operational phases to the development of the complete dataset, must be accompanied by continuous quality control (QC) and quality assurance (QA) activities.

In this chapter, we present the necessary steps and best practices for the QA/QC of micrometeorological data. These practices are vital to correctly identifying, addressing, and correcting errors that may occur during data collection, processing, and analysis. These processes ensure that the integrity and representativeness of meteorological datasets is maintained, despite challenges posed by instrumentation limitations, environmental conditions, and human errors. By standardizing methodologies and incorporating alternative data sources where appropriate, we aim to provide a framework for enhancing the reliability and applicability of micrometeorological data across a wide range of applications. This chapter serves as a comprehensive guide to both beginners and specialized audiences dedicated to improving the accuracy and usability of meteorological observations.

4.1 Data Quality: A Precondition for Any Application

Micrometeorological stations and sensors are installed to collect data for a variety of applications, including heat stress in urban environments, the microenvironment of crops and natural vegetation, the unique impacts of a given geographical site on atmospheric properties, or simply to obtain more detailed information about the local climate. Regardless of the purpose of the observations, the quality of the recorded data is determined by the accuracy of the measurements, the spatial and temporal representativeness of the records, and the representativeness with respect to the goal of the observation. With the increasing use of autonomous environmental sensors, particularly in low-power/low-cost applications, the risk of data loss and measurement error grows, emphasizing the importance of properly implementing quality planning and assurance in each step of the data collection process (Note 4.1, ◘ Fig. 4.1).

Note 4.1 Data quality

Core data quality attributes can be categorized as follows (Sturtevant et al. 2021):

- internal consistency, or the theoretical compatibility of the measured variable with related variables measured by the same instrument (e.g. temperature and dew point temperature, relative humidity, and precipitation)
- validity
- spatial/temporal representativeness
- completeness
- coverage
- availability
- traceability
- timeliness
- reliability (minimal data loss).

The goal of quality assurance (QA) and quality control (QC) is to detect and reduce errors in the measurement process. Quality assurance (QA) is the establishment of standards and procedures to ensure data continuity. Quality control (QC) is the set of methods that is used to ensure that the measurements meet established standards. Establishing a standard protocol for evaluating data informs users about the types of errors and limitations to expect during analysis.

In this chapter, we examine the various types of errors that may be present in measured data, their likely causes, and possible correction methods. Three QC procedures are typically performed:

- QC checks for possible errors in the data records, such as values far out of the range characterizing the climate at a given measurement site and season, or notably large discrepancies between values re-

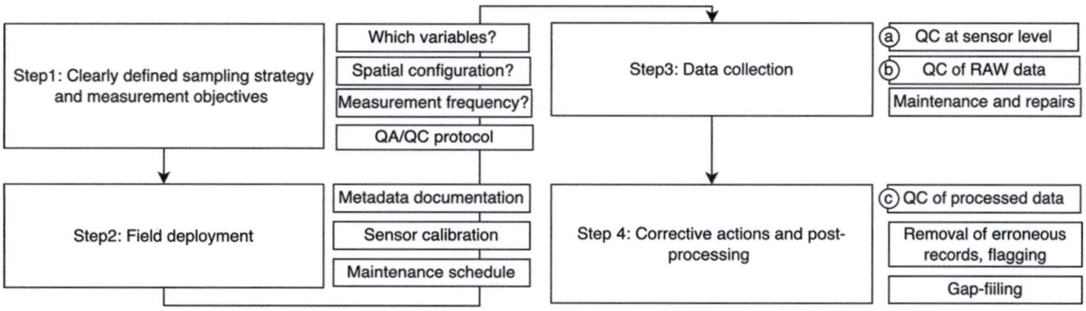

Fig. 4.1 Step-by-step data management and QA/QC procedure. Different QC techniques are needed for different parts of the data collection and processing chain: **a** QC at sensor level; **b** QC of RAW data; **c** QC of processed data

corded within close temporal or spatial intervals.

- Data gap checks for possible data missing due to instrument errors or other technical barriers.
- Checking for possible non-climatic changes (also known as inhomogeneities) in the time series of measured data, ensuring the integrity of the measurements. Non-climatic changes, i.e. inhomogeneities, may arise, from a variety of sources, including changes in instrumentation, installation practice, sheltering technique, and the meteorological station's microenvironment. One distinguishing feature between QC issues and inhomogeneities is that the former may affect isolated data, whereas the latter appears in sections of time series. It should be noted, however, that a QC procedure may detect quality problems not only for isolated data, but also for sections of time series, regardless of whether such issues are considered inhomogeneities or not. This also implies that QC procedures may detect inhomogeneities, but only if the inhomogeneity bias magnitudes are unusually large.

When a QC procedure detects incorrect data, the observation is flagged or marked as missing, unless the cause and severity of the error can be determined and corrected. Data gaps caused by incorrect observation or data recording, as well as missing observations, can be filled using spatial interpolation with data from nearby observing stations, or by modelling data fields. Finally, inhomogeneities are typically treated using statistical homogenization procedures. In every case, the original observational data is saved. Even erroneous data is typically saved to provide a trail of data post-processing when one or more steps of error correction by quality control, gap-filling, and homogenization are performed.

4.2 Sources of Errors in Observations

Perfect accuracy does not exist in experimental physics; measurement errors can be reduced by employing increasingly sophisticated observation techniques, but they will never be eliminated. The determination of whether an error is "small" is subjective and use-case-dependent, so we generally intend to reduce all types of errors that may affect the measured data. The measurement datasets contain various types of errors originating from various sources (Fig. 4.2). The majority of these errors are easily eliminated through proper quality control and comparison of measurement datasets. When erroneous values or inexplicable anomalies are discovered, identifying error sources may aid in the correction or reduction of biases, as well as lowering the risk of repeat error occurrences. The accurate and detailed documentation of the technical conditions of the observation in metadata greatly aids in the identification of the sources of observation errors and inhomogeneities.

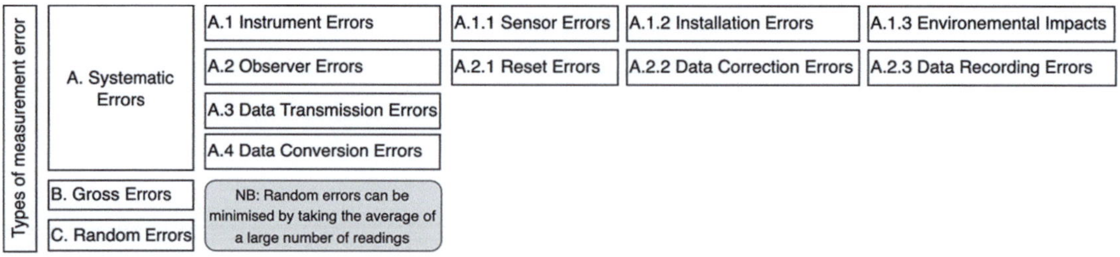

Fig. 4.2 Types of measurement errors

In Fig. 4.2, we outline the types of errors that can occur in measurement:
- gross errors
- systematic errors
- random errors.

Gross errors can arise due to human mistake, experimenter negligence, or equipment malfunction. Random errors are less easy to identify but can arise due to changes in experimental conditions (e.g. short/sudden changes in humidity or temperature within the sensor enclosure or signal attenuation) that are not related to environmental changes. These errors can be reduced by taking the average of a larger number of readings. Systematic errors can be grouped into instrument errors, observer errors, data transmission errors, and data conversion errors, and are discussed in more detail below.

4.2.1 Instrument Errors

When we talk about instrument errors, we often think of sensor errors, but sensor errors are just one type of instrument error. Depending on the instrument, errors may occur due to the following reasons:

Sensor errors: the maximum tolerable errors are determined for the most important instruments. Because manufacturers usually guarantee the maximum error, the occurring errors are controlled by instrument calibrations. In manual weather stations, small corrections are frequently applied for errors known from calibrations. The main technical instrument characteristics for quantifying the quality of observations are (Sturtevant et al. 2021):
- sensor accuracy
- precision
- sensitivity
- drift
- stability
- measurement range or span
- response time
- measuring lag.

Installation errors: Irregular installation will cause biases in the observed values, so strict adherence to instrument installation rules is essential. As installations age, they may lose their original characteristics, resulting in measurement errors. For example, a precipitation gauge placed on an inclined stand will not accurately detect precipitation, a grown nearby tree will obstruct wind measurement, and the shadow of a nearby column can pass over a radiation sensor or the shelter of an air temperature/relative humidity sensor every day during a similar time.

Lack of maintenance: The instruments and their installation must be kept clean and in the proper condition for climate observations. Measurement errors may occur as a result of irregular or poor maintenance.

Environmental impacts: When the sensor is exposed to the open air, contamination, weather conditions, or animals may interfere with the instrument's proper operation. Contamination can occur despite regular maintenance when smog, dust storms, or other factors cause unusually intense contamination. Weather can affect the accuracy of the instrument operation in a variety of ways, such as radiation, humidity, and snow-ice depositions on sensors, which can cause biases in the measured values. Small snow particles can penetrate partially closed places, such as Stevenson screens, and even with a Pitot tube, the intrusion of snow particles may halt wind gust observations. The list of weather effects could be expanded with numerous additional examples. Finally, animals (insects, birds) can affect micrometeorological observations, though they rarely have a significant impact on the observation of critical climate variables.

4.2.2 Observer Errors

Most meteorological observations are based on automated equipment, but exceptions do occur, and when observers must treat and read instruments, they occasionally make errors. The frequency and magnitude of observer errors vary greatly depending on the observed climatic variable, the conditions of observation, and the observers. Observer errors are typically smaller for instrumental observations than subjective observations. Natural conditions influence errors; for example, cloudiness observations are more accurate during the day than at night when there is no moonlight. Non-natural factors can also influence observer errors: artificial lights can affect cloud observations at night, while their absence can affect transparency observations. The observer must have adequate tools to complete the observations even in adverse weather conditions; otherwise, observation frequency and/or quality will be weather-dependent. One major advantage of observation automation is that it eliminates observer errors. The primary types of observer errors are listed below:

Instrument reading errors: When the observer's eyes are in an incorrect position, reading errors occur; however, for some instruments, it is also necessary to keep the instrument in the proper position for the correct reading (for example, liquid in glass minimum thermometers).

Instrument resetting error: Certain instruments, such as traditional liquid-in-glass thermometers used to measure minimum and maximum temperatures, require manual resetting. In the case of automatic weather stations (AWSs), regular maintenance or emergency service can lead to sensor resets. Since the data logger records signals from the sensors as long as it is powered on, it's important to power off the device, or at the very least, note the exact time the intervention starts and ends. This ensures that any erroneous measurements can be accurately identified and excluded from the data series during subsequent data processing.

Erroneous data correction: When readings are corrected for known or estimated instrument biases, such corrections must be applied consistently and clearly documented.

4.2.3 Data Recording and Transmission Errors

During data processing and transmission, errors can arise as a result of:
- signal noise—electronic noise which can interfere with the signal from the sensor
- data loss or corruption during transmission from the sensor or during processing
- errors in the algorithms used to process raw data into usable information
- data conversion errors.

The observed values can be recorded in a variety of measurement units. For example, precipitation can be measured in millimetres or inches, wind speed in metres per second, kilometres per hour, knots, and so on. When periods of observations recorded in different measurement units are combined without properly applying the necessary conversion, data conversion error occurs.

When only one or a few isolated pieces of data are impacted by errors, they are frequently hidden in natural variation. Quality control, on the other hand, can detect and eliminate errors of significant magnitude. Inhomogeneities can occur when errors affect long sections of a time series on a continuous or repeated basis.

4.3 Sources of Inhomogeneities in Observations

Any instrument, observer, data transmission, or data conversion error (▶ Sect. 4.2) can result in inhomogeneities if it occurs persistently or repeatedly over the course of the observations (systematic error). Inhomogeneities, however, can be caused by a variety of other factors. They are a result of changes in the technical conditions of the observations, and *the majority of them cannot be considered errors*, as the observed data are frequently correct both before and after the change, despite the fact that they include the effects of various technical conditions.

According to WMO's "Guidelines on Climate Metadata and Homogenization" (Aguilar et al. 2003), "most long-term climatological time series have been affected by a number of *non-climatic factors* that make these data unrepresentative of the actual climate variation occurring over time. These factors include changes in instruments, observing practices, station locations, formulae used to calculate means, and *station environment*". At this point, we would like to argue that station environment, i.e. land surface is an important element of the climate system and should not be considered as a source of inhomogeneities. However, to keep this book in line with concepts prevailing in the literature, we will include environmental changes in our analysis.

The following are some of the most common sources of inhomogeneity:

Change in Observation Time: Because automatic weather stations observe and record weather parameters in near-continuous fashion, changes in observation time are more likely to occur than in traditional, manned observations. The time period to which a recorded AWS value refers to may differ from previous standards. For example, the traditional period used in daily precipitation totals is from 07 h (local time) on the referred day to 07 h the following day. However, in many automatic weather stations, daily values are recorded for the hours 0–24 h of the referred day. Similar inconsistencies can occur when recording daily minimum and maximum temperatures. Such changes, if not properly recorded and corrected, may result in large-scale inhomogeneities.

Change in the Condition of the Instrument or Its Installation: The properties of a newly installed instrument will gradually change over time, though such changes are usually imperceptible for a long time. These changes necessitate the need for periodic calibrations. Simpler factors, such as the cleanliness, colour, and smoothness of the exteriors of the instruments and their shelters, can cause inhomogeneities. Unlike the previously mentioned inhomogeneity sources, such changes are not associated with a specific date but are continuous over time. For example, when the meteorological station and its instruments are cleaned, the observation's technical conditions can change abruptly. It should be noted that data inhomogeneities caused by periodic changes in dirtiness and cleaning are minimal and completely imperceptible in properly managed meteorological stations.

Changes in Land Cover Beneath and Around the Meteorological Station: These changes result in gradual changes in the technical conditions of observation. Such inhomogeneities are minimal in properly managed meteorological stations.

Environmental Changes in the Meteorological Station's Surrounding Area: Growing urbanization, changes in land use or the extent of irrigated area, and so on all contribute to gradually increasing inhomogeneities in comparison to the state prior to the changes.

4.4 General Aspects of Quality Management

The QA process and its principles are common to various types of environmental science measurements (Note 4.2), but how they are implemented varies depending on the measurement program (continuous or field campaign) and application goal.

> **Note 4.2** Quality management
> Field measurements, such as various micrometeorological measurement programs, are a three-step process that includes pre-field planning, field performance, and post-field documentation. These three phases can also be viewed from a data processing perspective, as illustrated in ◘ Fig. 4.3. The principles are the same; however, the implementation and specific tasks varies depending on the measurement program. Visits to measurement sites, data monitoring dynamics (daily, weekly, monthly, seasonal, etc.), and other tasks are determined by the specifics and requirements of the measurement program. Below is an example of a QA flow for continuously operating measuring sites.

Professional climate observations adhere to international standards throughout the data production process in order to provide high-quality observational datasets for climatologists and other users, and quality assurance is an integral part of the professional treatment of observed climatic data. Beginner users and specialized public users are also encouraged to follow or approach standard procedures in their practice, as long as the processes and protocols are economically feasible.

Sturtevant et al. (2021) define the quality management workflow as follows:

- defining the objectives and requirements
- planning the measurement system
- site setup and configuration
- calibration and testing
- inspection and maintenance
- data processing and management
- software-based and manual quality control
- performance monitoring and auditing.

Time series with significant quality issues or large magnitude inhomogeneities are either corrected or discarded based on the existence or absence of opportunities to provide equitable corrections.

Gap-filling and homogenization are recommended and frequently used steps in dataset development, but they are less regulated than the ones previously discussed. Datasets can be used with or without gap-filling and homogenization; exceptions include time series with visibly large inhomogeneities.

4.4.1 Best Practices for Meteorological Observations

One important goal of meteorological measurements for professional use is to provide data that is accurate and comparable over time. As a result, professional networks are designed so that meteorological station locations, instrumentation, and microenvironment remain consistent over time. On the other hand, in the case of agricultural and forest measurements, for example, the sole purpose of measurements is to determine impact of microenvironment changes on meteorological conditions. We believe that understanding standard meteorological observation processes and post-processing operations allows users to select near-optimal options for individual cases of meteorological station planning and operation, as well as the treatment of measured data.

A list of key general recommendations for operating meteorological stations are:

- Meteorological stations are located on a flat surface away from any high objects that could interfere with free ventilation of air from any direction, resulting in an undesired radiation surplus or deficiency in the meteorological instrument's microenvironment. Micrometeorological measurements intended to include

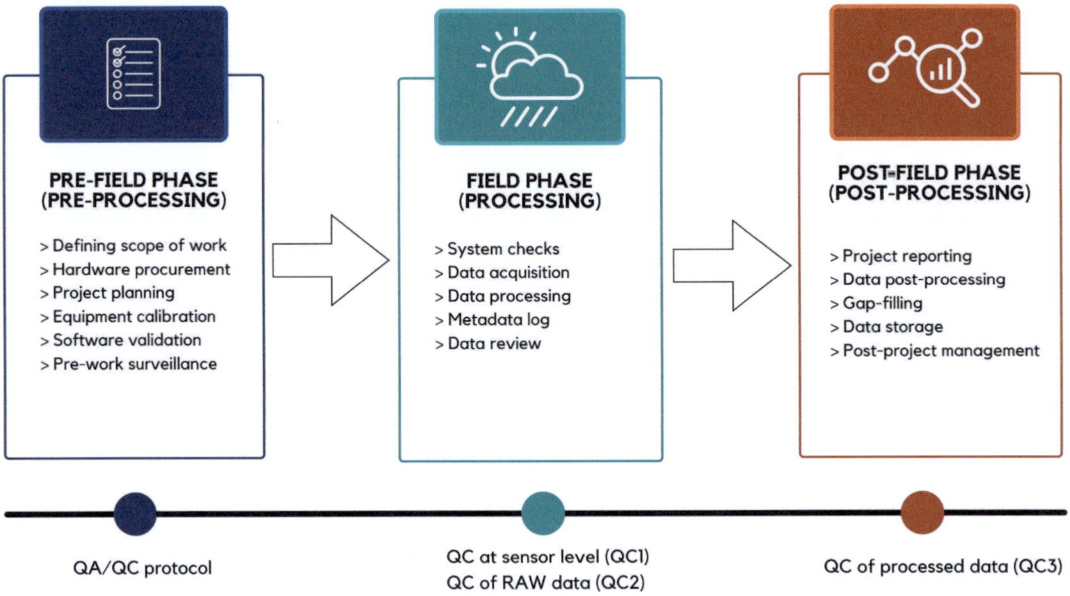

Fig. 4.3 Quality assurance (QA) tasks within a project typically follow a three-step process that includes pre-field planning, field performance, and post-field data-processing and project reporting

high objects (urban measurements, forest measurements) in their environment are exceptions to this rule.

- Meteorological stations are installed and operated exactly according to the rules described in previous chapters of this book.
- Basic station metadata includes the geographical conditions of the observing site, as well as the types and heights of meteorological instruments. Photographs of the meteorological station and its immediate surroundings are captured and saved.
- The meteorological station is regularly inspected and kept clean in the controlled microenvironment. Possible irregularities are recorded.
- Unnecessary changes to the meteorological station and its surroundings are avoided. When maintenance, repair, or instrument changes are required, they are completed and documented.
- When a station relocation or instrumentation changes are planned, it is best to perform parallel measurements for a few years (when feasible). In such parallel measurements, all but one of the technical conditions are the same for two meteorological stations, i.e. they may differ in the observing site, or the meteorological instrument used, but only in one, allowing the effects of a given technical change on the observed climatic characteristics to be quantified. Additional information on standard observation protocols and guidelines can be found in many WMO publications (2019, 2023a; b, 2024).

4.4.2 Metadata

Metadata, commonly called "data about data", provide information about data without revealing it (see also Appendix in Table 4.5). Regardless of whether the data will be used to extract knowledge or for modelling, it is important to know as much as possible about how and where measurement is conducted and under which circumstances. Therefore, metadata can be classified according to Zeng and Qin (2020) as follows:

Descriptive metadata refer to full set of metadata associated with particular (micro)meteorological measurements and fully described in "Guidelines on Climate Metadata and Homogenization" (Aguilar et al. 2003).

Structural metadata are metadata about containers of data: file type, format, size, digital object identifier (DOI), or some other digital identifier.

Statistical metadata contain information about statistical characteristics of data including results of quality control and gap-filling (list of identified red flags for both).

Reference metadata are metadata about the methods used in gathering metadata (observations, measurements, data bases including satellite observations), performed statistical analysis, quality control, and gap-filling.

Administrative and legal metadata provide information about permissions and public licensing, and creator and copyright holder.

Typically, DOI associated with data always includes the above listed metadata. Metadata categories used by the WMO WIGOS system can be found in the WMO reports (2019, 2023a).

When measurements are being collected at multiple sampling points, sometimes by different partners within the same project, properly documented metadata supports measurement repeatability and comparability. Below we give an example of basic metadata categories, ◘ Table 4.1.

4.5 Methods of Data QC

The main goal of data quality control (QC) is to determine whether the data meet overall quality goals and defined quality criteria for individual values. This section focuses on the QC methodology and strategies for preventing errors from entering the dataset, ensuring data quality during collection, and maintaining data quality throughout the project. QC is essential for detecting, flagging, and addressing errors in measurements, ensuring data reliability for future applications. Key indicators of data quality problems, such as

◘ **Table 4.1** Example metadata sheet following Brunet et al. (2020)

Category	Metadata type
Station/platform identifiers	Abbreviated sampling location identifier
	Begin date/hour
	End date/hour
	Type of weather station/logger
	Responsible organization
	Variables being recorded and their measurement units
	Measurement frequency
	Time zone
Geographical data	Latitude
	Longitude
	Elevation
	Topographical information
Local environment	Local land use/land cover
	Obstacles, skyline survey, site photographs from North, South, East, and West
	Soil type
Instrument maintenance	Sensor type and accompanying data sheet
	Radiation shield and mounting
	Calibration and maintenance schedule

out-of-range values, spatial inconsistencies, and logical contradictions, are identified and discussed. Techniques ranging from plausibility tests to higher order statistical analyses are outlined, providing a structured approach to error detection and correction. The importance of adapting QC methods to specific applications and datasets is emphasized, ensuring that the integrity of the data is maintained throughout its use.

4.5.1 Indicators of Quality Problems

The purpose of quality control (QC) is to detect and filter out erroneous observational records. Very small errors are sometimes undetectable, whereas large errors are easier to detect and remove. The indicators pointing to large data errors are briefly presented here, and they will be discussed below:

- Occurrences of values out of the climatologically expected range: Unusually high or unusually low values may indicate either true climatic extremes or erroneous data recording. The distinction between these two options needs particular attention in quality control procedures.
- Sudden changes in the recorded values of continuously changing weather parameters like temperature, humidity, or atmospheric pressure: Such sudden changes may indicate the passage of a sharp weather front or erroneous data recording. The distinction between these two needs to be checked, if a weather front truly passed through at the time of the sudden change. Another indicator of a data sequence issue are flat lines, i.e. repeated recording of a given value. Sometimes the weather parameters do not change for several hours, but the occurrences of unusually long or unusually frequent flat lines are strong indicators of instrument errors.
- Spatial inconsistency: Large deviations from the data of nearby weather observations may occur for a local weather event like local fogs or thunderstorms, but in other cases they indicate erroneous data recordings.
- Logical errors: When a group of synchronously recorded weather data cannot jointly occur, such incidents point to a data recording error or errors. For instance, when a meteorological station reports snowfall and higher than 7 °C air temperature at the same time, at least one of the observed parameters is erroneous.
- Unexpected frequency distribution: The frequency distribution of the observed values of a weather parameter in a given geographical site and season of the year can be estimated based on general climatic knowledge. For instance, the frequency distribution of land surface air temperature is a nearly symmetric distribution with much frequent occurrences of near average values than extreme low and extreme high values. When the frequency distribution of the recorded data shows large deviations from the expected distribution, this may indicate a long-standing error in the operation of the instrument or an error in another phase of the data recording.

A summary of key QC tests used for identifying errors in time series data is provided in Table 4.2.

Out-of-Expected Range Values: When unusually high or unusually low values are recorded, they may indicate instrument error or erroneous data recording, although extreme weather events may also produce the occurrences of unusual values of meteorological elements. One task of QC procedures is to decide if an out-of-expected range value is a true occurrence of an extreme climatic event, or if it occurred due to a technical error of the observation. Erroneous out-of-expected range values are called outliers, while accepted out-of-expected range values are called extreme values. We can distinguish three types of out-of-expected range values:

Table 4.2 Overview of characteristic QC tests for error identification in observed time series data (Taylor and Loescher 2013; Sturtevant et al. 2021)

QC test category	Purpose/problem	QC tests	References
Plausibility tests to ensure only electronically and atmospherically feasible values are retained	Detect out-of-range values	Range test	Taylor and Loescher (2013)
	Change in variance structure	Sigma test	Taylor and Loescher (2013)
	Data stuck at a singular value	Delta test	Taylor and Loescher (2013)
	Jumps in data values	Step test	Taylor and Loescher (2013)
	Missing data point	Null test	Taylor and Loescher (2013)
	Multiple missing data points	Gap test based on sampling frequency	Taylor and Loescher (2013)
Spike tests	Outlier removal	Window tests, filter-based tests	Brock (1986), Metzger et al. (2012), Starkenburg et al. (2016)
Higher order moments of statistical distribution	Check the statistical plausibility of the values	Skewness, kurtosis	Vickers and Mahrt (1997), Finkelstein and Sims (2001)
Comparison with collocated/independent data	Check data consistency	Climatological and temporal check, spatial consistency	Pastorello et al. (2014), WMO (2024)

- **Physically Impossible Values**: The physically possible values of weather parameters have finite ranges. These threshold values are sometimes derived directly from the definition of the weather parameter, while in other cases they are linked to observed extreme values. For example, relative air humidity is defined as the percentage of actual humidity in relation to the maximum humidity that air at a given temperature can contain when saturated. As a result, the only physically valid values for relative humidity are 0–100%. Similarly, wind speed, gust, and precipitation amount cannot have physically valid values less than zero.

In Fig. 4.4, a coding error resulted in wind directions that were not within the expected range of 0–360°. Fortunately, by correcting the method and reprocessing the data, processing errors can be reduced. It is critical to preserving the raw data in a format that closely resembles the original sensor values, as computational errors of this nature are common (Sturtevant et al. 2021).

Additional threshold values for physically valid ranges of weather parameters are listed in Table 4.3.

While new absolute records should not be ruled out, this possibility is relevant only in regions where values close to the observed absolute extreme may plausibly occur. For example, when looking at extreme minimum temperatures in the Antarctic, maximum temperatures in the world's hottest deserts, and extreme high precipitation amounts in the Himalaya, one should consider the possibility of a new record being set. However, excluding such extreme examples, we can conclude that the detection of a value outside of the ranges shown in Table 4.3 indicates a

4

certain outlier value, also known as a physical outlier:

- **Climatically Impossible Values**: In a given geographic region and for a specific season of the year, the ranges of weather parameter values are typically narrower than the range of physically possible values. Based on an observing site's local climate, site-specific ranges of climatically possible values can be calculated with caution to avoid excluding any true extremes. For example, a recorded negative temperature value in summer along the Mediterranean coast is undoubtedly an outlier value because it is climatically impossible.
- **Climatically Unlikely Values**: A climatically unlikely value may indicate the true occurrence of an extreme climatic event, or an erroneous recording. They are suspicious values, and the decision if any of them is a piece of erroneous data requires a detailed examination of their occurrences. In a QC procedure, the climatically unlikely values are chosen by performing simple statistical calculations that separate climatically unlikely values from other data. A very simplified procedure to identify climatically unlikely values can use daily climate extremes from the nearest climate station and include expected effect of climate change typical for the season and the region.

Table 4.3 Example of physically acceptable range limits of weather parameters for instantiations measurements. These limits are dependent on the season and the conditions of the measuring instrument. They can be configured as broad and general at this point in the QC process and adjusted if necessary to reflect the climatic conditions of the region more precisely (Petrić et al. 2020; WMO 2023b)

Variable	Minimum	Maximum
Air temperature	– 80 °C	60 °C
Relative humidity	0 %	100 %
Dew point temperature	– 80 °C	35 °C
Precipitation (1-min interval)	0 mm	40 mm
Wind direction	0°	360°

Suspicious values should be examined one-by-one if their occurrences are related to an extreme weather event. During such examinations, the data of other climatic variables observed at the same observing station, the observed climatic data of the neighbouring observing stations, and the characteristics of large-scale circulation should be taken into account.

Data Sequence Errors: Three types of data sequence errors can occur in the time series of observed climatic data:

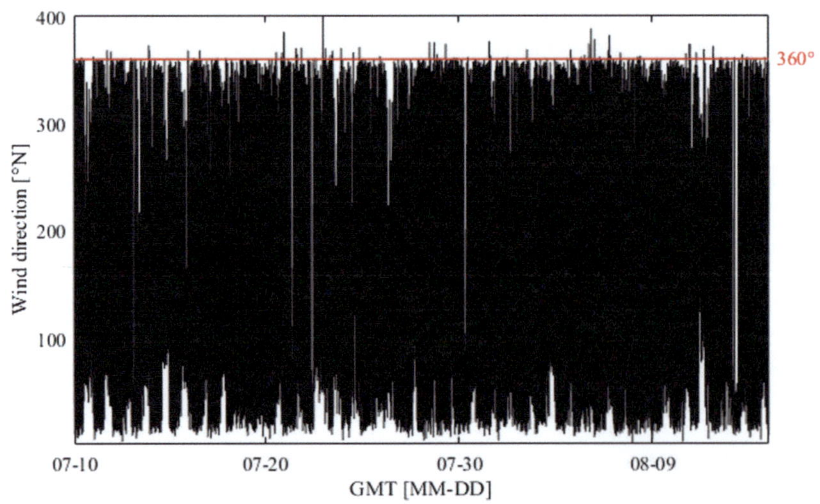

Fig. 4.4 Erroneously calculated wind direction data based on measured vector components of a 2-D sonic anemometer; note that some of the data values are beyond the theoretical limit of 360 N. After Sturtevant et al. (2021), with kind permission of Springer

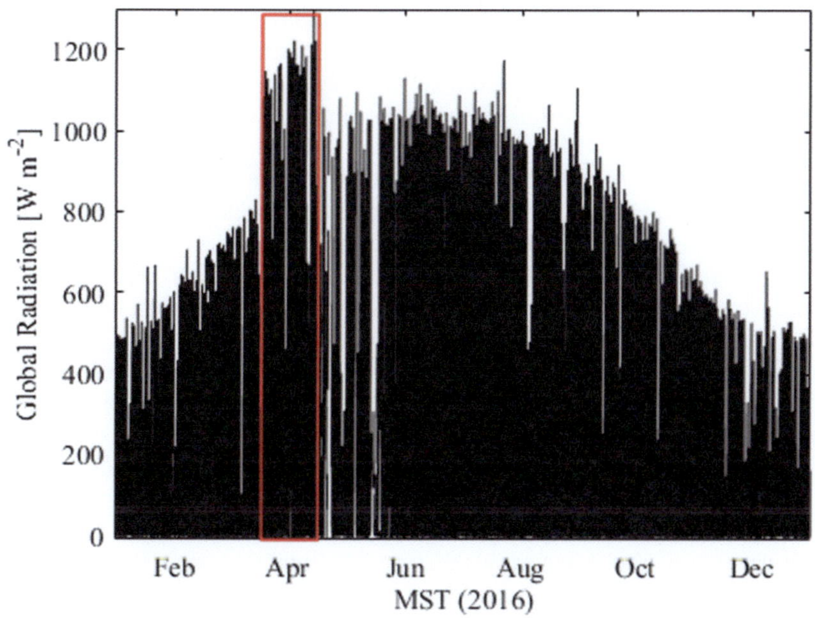

Fig. 4.5 Level shift in global radiation measurement (outlined in red) caused by changes in sensor calibration (MST—Mountain Standard Time). After Sturtevant et al. (2021), with kind permission of Springer

- Repeated recordings of the same value in long streaks, also known as flat lines
- Scattering of values with unrealistically large amplitude
- Sudden sharp changes in values.

Some weather parameters change slowly over time (temperature, atmospheric pressure, air humidity), while others change quickly (wind speed, precipitation intensity). While the first two error types can occur with any weather parameter, sharp change occurrences only indicate potential errors in a time series of gradually changing weather parameters:

- **Flat lines**: Streaks of repeated values in weather data often signal technical errors but can occur naturally for certain meteorological elements. For instance, it's normal that recorded precipitation or leaf wetness are zero over long periods. In the time series for daily mean or extreme values—like temperature, humidity, and wind speed—a few days of repetitive data are expected, but if it takes longer than three days it typically suggests errors. These errors are likely from automated systems in observation and data recording. Furthermore, rounding measured values is not acceptable.
- **Excessive scattering of data**: Unrealistic scattering is a clear indication of a technical error in which the distribution of the recorded data shows large deviations from the expected frequency distribution.
- **Occurrences of unusually steep changes in the data series**: A sudden shift in the observed data levels (Fig. 4.5) could be caused by a change in the technical conditions of the observation, implying that either before or after the shift, some error affected the observation or recording, which can be due to changes in sensor calibration, or temporary operational failure. It should be noted that powerful weather fronts can cause rapid, large, and persistent changes in atmospheric conditions; thus, all flagged events require manual control.

Spatial inconsistency occurs when there is a significant difference between the observed value of a meteorological element at a specific site and the values recorded at nearby stations at the same time. Such observed values are suspicious, even if they are within the climatologically expected range. The thresholds for acceptable differences between neighbouring stations are determined by the specific meteorological element being measured, the distance between the stations, and the spatial density of the observations. More details about how to set appropriate thresholds using neighbouring data can be found in Taylor and Loescher (2013).

For any meteorological variable, the data from neighbouring stations must be to-an-extent correlated; otherwise, large spatial differences occur at random. However, even in spatially dense observing networks, some occurrence of large spatial differences cannot be excluded when resulting from significant local weather events such as local showers, thunderstorms, and fog.

Theoretical inconsistency: Daily maximum temperatures cannot be lower than daily minimum temperatures on the same day; the maximum wind gust of a given period cannot be lower than the mean wind speed of a given period; and snow depth cannot increase between two consecutive recordings if no precipitation fall occurred between the two recordings of snow depth values. When daily mean temperatures are reported alongside other observed data, they cannot be lower than the daily minimum temperatures or higher than the daily maximum temperatures observed on the same day.

Theoretical inconsistencies may indicate serious problems with observation or data recording, but minor errors in data post-processing can also cause logical inconsistencies. For example, when positive corrections are applied to observed minimum temperature data, the corrected minimum temperature exceeds the observed maximum temperature by a few tenths of a degree on a day with little temperature variation (due to rain, fog, or other factors). The distinction between serious problems with observation and minor errors in data processing necessitates an individual examination of the occurrences of theoretical contradictions. However, logical inconsistencies are always considered errors and must be corrected.

Unexpected frequency distribution: Most QC procedures control the reliability of individual values, with the exception of controlling the frequency distribution of observed values. We know the general shapes of the probability distribution of weather elements from climatological studies, at least roughly, and the frequency distribution of observed values should be similar. Temperature and atmospheric pressure, for example, have nearly Gaussian distributions with some known differences depending on geographical regions and seasons of the year, precipitation amounts have nearly exponential distributions, wind speed has a strongly asymmetric distribution with maximums at low values (positively skewed distribution), and so on. The control of the frequency distribution necessitates knowledge of the climate in the studied area. When large differences appear in comparison to the climatologically expected distribution, it indicates that there are significant errors in many of the recorded data points.

Following error types can affect frequency distribution:

- Instrument and data logger errors
- Changes in the measurement units
- Erroneous merging of time series.

Meteorological instrument measurements are converted into standard physical units for the climatic variables being monitored. In automatic meteorological stations, the data logger's software performs such transformations. In modern climate observations, data are rarely transformed manually, but we must be cautious with any human intervention to the data, as such interventions have the potential to introduce errors. More importantly, one should check the correct recording of observed data on a regular basis, as well as control for any unusual changes in the statistical properties of the data caused by data logger or software changes. Merging time-series data produced by different meteorological stations may result in data incompatibility due to differences in measurement units. Temperatures can be measured in Celsius or Fahrenheit, precipitation in millimetres or inches, and wind speeds and gusts in kilometres per hour, metres per second, or knots. Differences may also occur in the coding of wind directions. Wind directions can be expressed as initials (N, NE, E, SE) or degrees in a circle, with 0° representing the northern direction and NE = 45°, E = 90°, SE = 135°, and N = 360°. Wind directions are sometimes rounded to the nearest 10° and expressed by the two first digits of the number of degrees. For example, NE=05, E=09, SE=14, and N=36. All these examples demonstrate the importance of exercising caution when combining time series collected at different locations and time periods. Another data transformation issue is the conversion of measured atmospheric data to the standard level, which is the sea level (except for high mountain stations). This conversion is based on physical rules (so-called barometric formula) and uses the height of the barometer over the sea level and the air temperature that is observed simultaneously with the atmospheric pressure measurement. Software error, or errors in the measurement of the height of the barometer, might cause erroneous data conversion.

Time shifts, or changes in data patterns over time relative to theoretical predictions, are a common but simple correction issue in meteorological measurements. Plotting radiation readings by time of day in local standard time, as shown in ◘ Fig. 4.6, reveals this inaccuracy, as the long-term diel average is expected to peak around midday. This error is caused by incorrect time settings in internal instrument computers or data loggers, as well as the incorrect use of a time zone offset during processing. However, detecting this inaccuracy is more difficult when the data lacks the expected temporal pattern. In this case, it would be appropriate to check the measurement timestamps in real time while collecting data. The offset between the actual and planned time zones of measurement must be determined and applied to compensate for this inaccuracy. The only way to avoid this issue is to ensure that the processing methods and instrumentation time settings are accurate.

Infrastructure interference is a dangerous error in atmospheric measurements that must be avoided through proper site planning and setup and detected by closely inspecting the data. Avoiding this mistake is critical because it is often difficult to correct after the fact.

◘ Figure 4.6 depicts an example of infrastructure that shadows the sensor throughout the afternoon, demonstrating how it may interfere with global radiation readings. Only when the effect's timing, pattern, and amplitude are known can this type of inaccuracy be compensated for. If not, the impacted data must be deleted or flagged as suspicious. Even better, data collection tests should be performed prior to the experiment as part of proper site design and configuration to ensure that the measuring platform does not interfere with measurements.

An understorey photosynthetically active radiation (PAR) sensor was covered in soil as a result of firefighting efforts, as shown

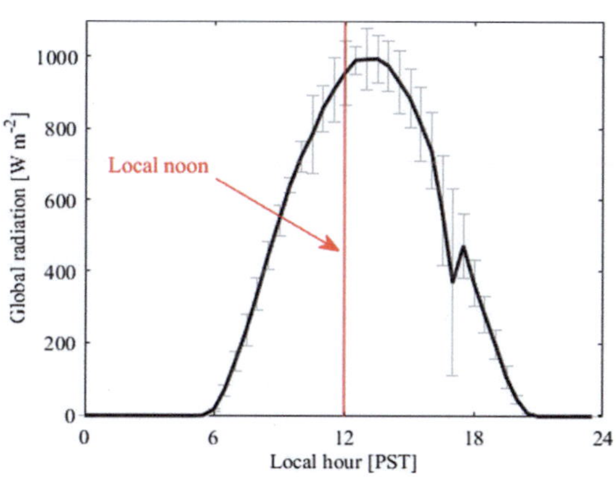

◘ **Fig. 4.6** Approximately one month of diel average global radiation measurement at Twitchel Island, California (CA, PST—Pacific Standard Time), showing improper time zone application as well as shadowing of the sensor by the site infrastructure in the afternoon. Error bars indicated standard deviations. After Sturtevant et al. (2021), with kind permission of Springer

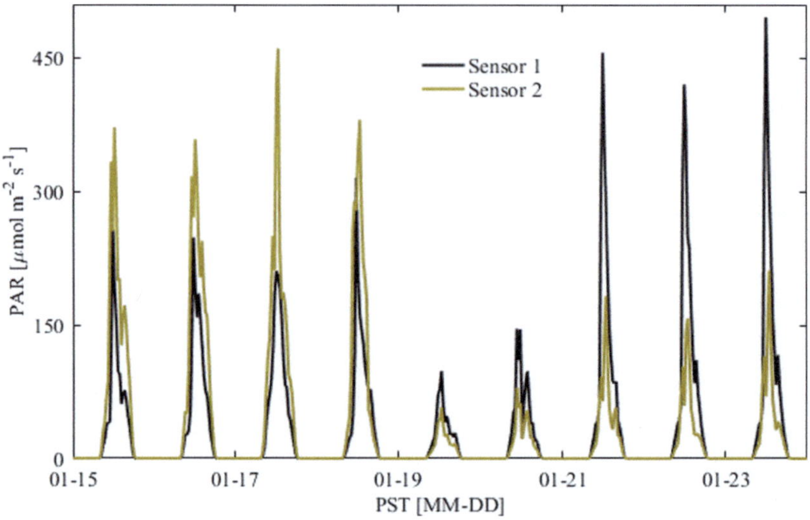

Fig. 4.7 Change in the relationship between two ground-based PAR sensors after sensor 2 was covered in dirt by fire-prevention activities on 18 January, After Sturtevant et al. (2021), with kind permission of Springer

in Fig. 4.7, which resulted in lower observed values. To reduce the amount of data affected by field interference, conditions must be measured carefully and frequently. These errors can originate from a variety of sources, are difficult to avoid, and are frequently overlooked by automated quality assessments. Detailed maintenance logs and site photos can help to reduce the amount of time spent determining which data to discard or mark as questionable.

4.5.2 Detection of Inhomogeneities

In the time series of observed data, the impacts of non-climatic changes (inhomogeneities) are mixed with true climate variation, and the real-time use of the data generally does not need the separation of these. Therefore, non-professional users should not homogenize data, except when they see large sudden shifts (breaks) in the series, which cannot be produced by climate variation. Even in such cases, it is better to use spatial differences of time series (relative homogenization) than analysing directly a single series (absolute homogenization). The inclusion of statistical break detection methods as supplementary tools to metadata use and visual inspections is recommended. The most popular statistical break detection methods are t-test, single break detection methods, standard normal homogeneity test, analysis of accumulated anomalies, and fitting of optimal step function. For each method, there are available open-source software libraries in R and Python, well equipped with documentation and follow-up literature (R Core Team 2024).

When statistical tests are used solely to control the statistical significance of individual break positions, the t-test is recommended. Note, however, that in professional homogenization more sophisticated software packages are used (Venema et al. 2020; Guijarro et al. 2023), which take into consideration the combined impacts of multiple inhomogeneities.

4.6 Methods of Data Correction, Homogenization, and Gap-Filling

Data failing QC are typically removed from the data series and categorized as missing data. When inhomogeneities affect a substantial part of the series, they should not be removed but addressed through homogenization techniques (Aguilar et al. 2003; Domonkos et al. 2023).

An important issue in homogenization and gap-filling (GF), which can significantly influence the choice of methodology, is related to the period after measurements and/or observation have been completed, within which the process must be finished. For operational purposes, this is typically one day. Scientific meteorological or operational climatological measurements differ significantly from operational measurements for non-climatological purposes. While the former can analyse data at the end of an experiment or at the scheduled time for publishing micrometeorological data, the latter requires almost immediate QC analysis and GF. As can be seen from Fig. 4.8, the significance of timely and efficient meteorological data management, QC, and gap-filling is evident ("Bis dat qui cito dat").

4.6.1 Gaps in Meteorological Data Time Series

Sources of Gaps in meteorological data time series can result from data disparity, misinterpretation of data, and missing data (including data removed after QC). Data disparity refers to the differences and/or discrepancies in the quality, quantity, or accuracy of data from different sources. Disparity of meteorological data can arise from various sources:

- Instrument failure: Data dropouts, or periods with no recorded data, most commonly occur due to technical malfunctions.
- Data collection method: Automatic weather stations (AWS), measurements made by humans, and remote sensing are typical meteorological data sources, all bringing completely different measurement methodologies. All methodologies are acceptable, but it is important to remember the differences among methods when using and comparing meteorological data from different sources.
- Measurement errors are a broad topic already addressed in ► Sect. 4.2.
- Inconsistencies in recorded data: AWS commonly record data with approximately 1 Hz frequency; according to users' preferences, the averaging interval is set to a period of interest (1 min, 1 h, etc.). Measurements performed by humans indicate values of meteorological elements at the moment of measurements. Therefore, while using and comparing data from AWS and provided by humans, significant differences should be expected between, for example, hourly averaged and current values of meteorological elements.
- Different data units and terminology: Regarding units, the most common example is the difference between temperature measured in Fahrenheit and Celsius. However, differences between the two scales are significant enough that mixing data is relatively hard, and that kind of error can be easily identified and removed. A bigger problem is when just the term "humidity" of air is used as a variable describing air characteristics. If units are not clearly stated, it is difficult to decipher if data are related to relative, absolute, or specific humidity. A common terminology problem is associated with using the term "climatological" data (30-year averages) and anomalies. If not stated differently, the anomaly is the deviation of measured data from the climate average, i.e. the normal value.

Misinterpretation of data typically originates from data aggregation, data variability, insufficient domain expertise, and statistical analysis conducted without checking the physical or climatological context ("Correlation does not imply causation.").

Aggregation of data is a technique that has recently become very popular due to significant attempts by the artificial intelligence (AI) community to test their methods on measured and simulated meteorological data and their need for as much data as possible. Data aggregation typically includes spatial aggregation (same measurement methodology at different locations), aggregation of different data sources (same location, different data sources for different periods, for example), and aggregation on both methodological and spatial levels. This type of aggregation can be a source of so many inconsistencies and misinterpretations that it should be avoided as much as possible when deal-

4

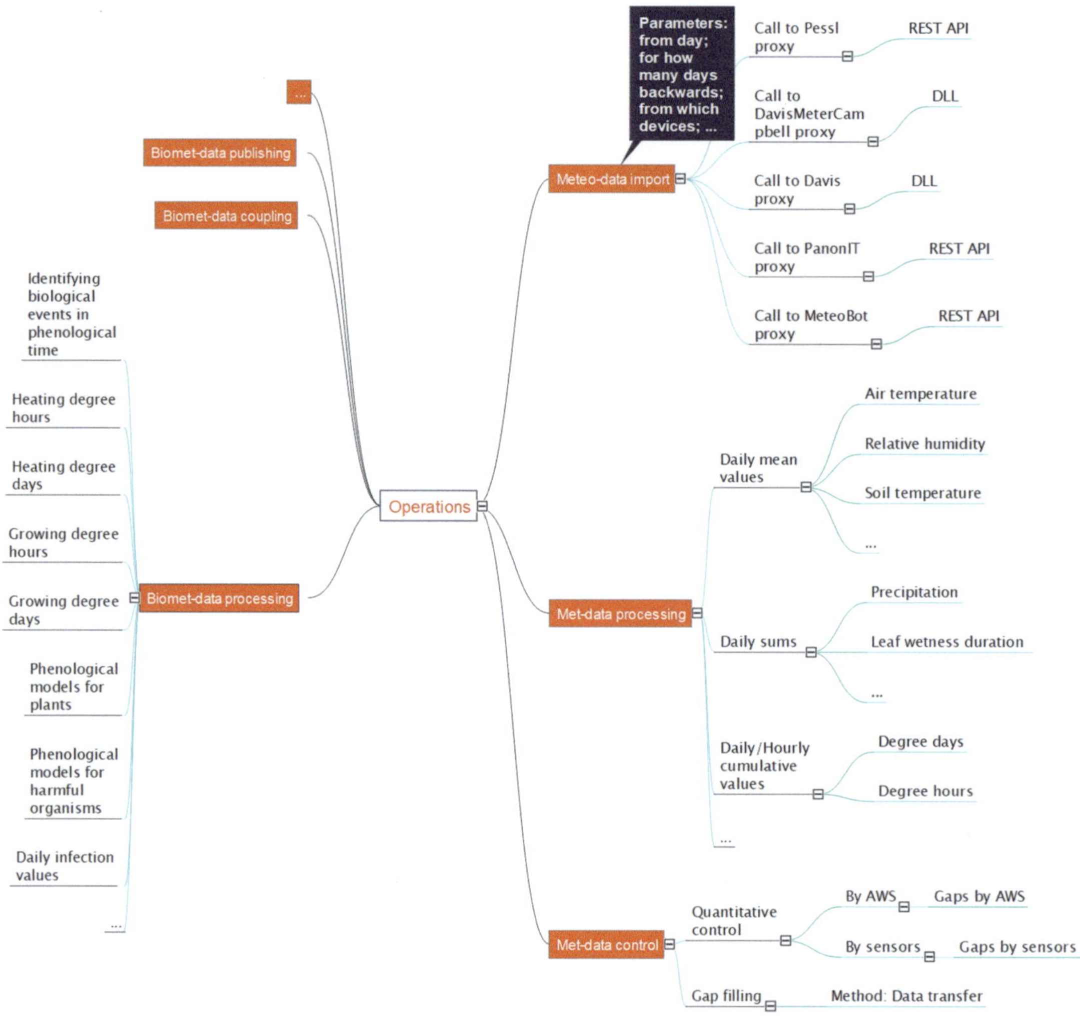

Fig. 4.8 Meteorological data-related segment of PIS (Serbia) data flow. (AWS—Automatic Weather Station; API—Application Programming Interface; DLL—Dynamic Link Library)

ing with meteorological data and very clearly marked if any analysis is based on aggregated data.

Variability is a common characteristic of meteorological data series. Even though it is difficult to define an upper threshold for a particular meteorological element, one criterion can be based on the variability of recent climatology. Variability higher than the climatological average can be related to disparity or less probable, but important events affecting micro-scale atmospheric conditions.

Possible Gap distribution analysis is an important step before selecting and applying a gap-filling methodology. Specifically, the spectral analysis of gaps and the duration of the highest frequency associated with a gap should be key factors in choosing a gap-filling method. For instance, if most gaps in an hourly data series last up to four hours, then simple interpolation will likely provide sufficiently good results (Fig. 4.9). However, filling gaps lasting several days should be performed using some proven methods, which will be detailed in ► Sect. 4.6.2.

Examples of gap distributions in meteorological time series obtained from a rural and an urban AWS network are presented in Fig. 4.9a, b.

Fig. 4.9 a–b Gap distribution in **a** rural (PIS, Serbia) and **b** urban (Ghent, Belgium) meteorological time series data from AWS networks

4.6.2 Filling Gaps in Data Time Series

A common classification of gap-filling methods includes statistical techniques, dynamic (physics-based) methods, machine learning (ML) methods, and hybrid approaches. Initially, the hybrid approaches combined dynamical and statistical techniques (see, for example, Lompar et al. 2019). More recently, the development of physics-guided neural networks (PgNN) has introduced a new concept and provided opportunities for further enhancements in gap-filling methodologies.

In selecting the gap-filling method, important criteria include the scale (duration) of the gap and the volume of measured or observed data at a particular location. Specifically, for ML methods to be used with acceptable accuracy for synthesized data, the training data series should include several seasons at the location of interest, or at least at surrounding representative locations.

Unfortunately, not all meteorological elements are equally subjected to gap-filling methodologies. While for air temperature different gap-filling approaches can be found in many studies spanning a wide array of techniques, for precipitation this number is reduced, and for all other meteorological elements it is hard to find standardized methods. Most frequently, they are methods developed for air temperature which are then applied to other meteorological elements (Körner et al. 2018; El Hachimi et al. 2023). In other words, the probability of gaps for all measured variables in a particular segment of the data series, especially if they are based on automatic measurements and caused by systematic instrument errors, is high (■ Fig. 4.9), but these gaps cannot be treated with the same techniques.

Filling gaps in air temperature data series: There are several approaches based on various statistical techniques that use historical data, in situ observations, and objective analysis for spatial interpolation of data to close gaps in near-surface air temperature data (up to 2 m). These gap-filling techniques can be classified into two types: temporal (Von Storch and Zwiers 1999; Beckers and Rixen 2003; Claridge and Chen 2006; Liston and Elder 2006; Stöckli et al. 2008; Blyth et al. 2010; Papale 2012; Vuichard and Papale 2015; Lompar et al. 2019; Cerlini et al. 2020; Dumitrescu et al. 2020; Dyukarev 2023; Jacobs et al. 2024) and spatial (Garen et al. 1994; Dodson and Marks 1997; Hartkamp et al. 1999; Daly et al. 2000, 2002; Rolland 2003; Stahl et al. 2006; Pape et al. 2009; Minder et al. 2010; Tobin et al. 2011). Interpolation methods such as inverse-distance weighting (IDW) (Daly et al. 2000), kriging and cokriging (Garen et al. 1994; Hartkamp et al. 1999; Tobin et al. 2011), multiple regressions (Stahl et al. 2006), and thin-plate splines (Pape et al. 2009) are commonly used for filling spatial gaps in temperature data. A minimum number of measurement stations is required in gap-filling approaches that rely on accurately representing the local observed lapse rate (Dodson and Marks 1997; Daly et al. 2002; Rolland 2003; Minder et al. 2010). In contrast, temporal gap-filling techniques are dependent on the autocorrelation of meteorological time series. Henn et al. (2013) provide a comprehensive comparison of both spatial and temporal gap-filling methods for hourly near-surface temperature including methods such as Empirical Orthogonal Functions (EOF), linear interpolation considering the diurnal temperature cycle, and three variations of lapse rate-based gap-filling. Lompar et al. (2019) developed a novel technique for gap-filling that is based on a combined dynamical-statistical approach. The obtained results indicate that ERA5 data can be used to fill temperature gaps. To obtain meaningful temperature data, however, ERA5 data must be debiased using measured data before and after gap. Other methods that examine the feasibility of using debiased ERA5 to fill gaps in temperature time series are explored in Vuichard and Papale (2015), Cerlini et al. (2020), Dumitrescu et al. (2020), Dyukarev (2023), and Jacobs et al. (2024).

Filling gaps in precipitation data series: Precipitation is one the most complex meteorological elements to analyse and forecast. Its spatial and temporal variability makes any gap-filling technique highly uncertain. Fagandini et al. (2024) present an overview of precipitation gap-filling techniques. The first method of this kind was the polygon method (Thiessen 1911) which was followed, many years later, by the natural neighbour method (Sibson 1981), the nearest neighbour method (Brandsma and Buishand 1998), and inverse distance weighting (IDW) which was developed by the U.S. National Weather Service in 1972. The common precipitation gap-filling method in climate analysis studies is the so-called “FAO method”. The FAO method addresses missing data in precipitation time series by using measurements from the period of the gap at nearby stations that show a strong correlation with the station of interest. Recently, numerous ML methods were tested in order to provide fine downscaling of gap-filling of precipitation with limited success and for selected locations.

4.7 Computer Techniques of QC and GF: MetObs Python Toolkit

MetObs (Vergauwen et al. 2024) is an open-source Python toolkit designed to assist both beginners and professionals conducting micrometeorological measurements, to overcome QC and GF issues and to facilitate the exploitation of their valuable datasets. MetObs offers a comprehensive framework that supports the entire workflow, from raw sensor data to dedicated analysis, and is adaptable to various types of non-traditional networks without encountering formatting issues (◘ Fig. 4.10).

The starting point is always importing a raw data file into a MetObs dataset, which is done by a template file. This file contains all the information on how to interpret the raw data (e.g. data structure, time zone, units, sensor description). A built-in dialog assistant will create the template file by asking questions (◘ Fig. 4.11).

Once the raw data is imported into a MetObs dataset, the time resolution is firstly resampled to the desired resolution, followed by identifying erroneous and missing records (◘ Fig. 4.12). Dedicated software packages for quality control, such as TITAN and CrowdQC+, already existed prior to the development of MetObs and are therefore implemented in the toolkit. For commonly used quality control checks, a Python version is implemented in the toolkit, to simplify the installation process for the users.

Finally, missing records are filled in with the most suitable or preferred gap-filling method. Both interpolation and debiased-model data gap-filling methods are implemented in the toolkit (◘ Fig. 4.13). The latter uses ERA5 data for the location of the corresponding stations, which is directly transferred using the Google Earth Engine and cloud infrastructure. These ERA5 time series are then bias-corrected by using the observations that passed the quality control and are used to fill the gaps.

At this stage in a typical workflow, a continuous, quality-assured dataset is obtained, and ready for analysis. The MetObs toolkit is equipped with standard analysis tools and visualizations such as filter methods, diurnal cycles, seasonal cycles, and land cover correlations (◘ Fig. 4.14a, b).

Using the Google Earth Engine, metadata of the location can be extracted for all stations (◘ Fig. 4.15).

MetObs was developed in such a way that people without a coding background can utilize it to get insight into their own meteorological measurements by following examples and using tutorials. At the same time, it allows more experienced data scientists to tweak the functionalities that the toolkit provides, a pipeline for their dedicated use case.

The MetObs toolkit is published by Vergauwen et al. (2024). If you are interested in using the MetObs toolkit, the documentation is the best start point: ► https://metobs-toolkit.readthedocs.io/en/latest/index.html, access 13 Nov. 2024. Feedback, requests, and collaboration comments can be posted on the official GitHub page: ► https://github.com/vergauwenthomas/MetObs_toolkit, access 13 Nov. 2024.

4.8 Use of Already Available Alternative Data Sources

Alternative available data sources for micrometeorological applications can normally not fulfil the same quality standards as an in situ measurement, which are needed for many specific applications. They can serve however as a control parameter for station failures, potential gap-filling, or as a low-cost supplementary information.

Different (micro)meteorological data of *past and real-time nature* from different measurement standards could be in general available from:

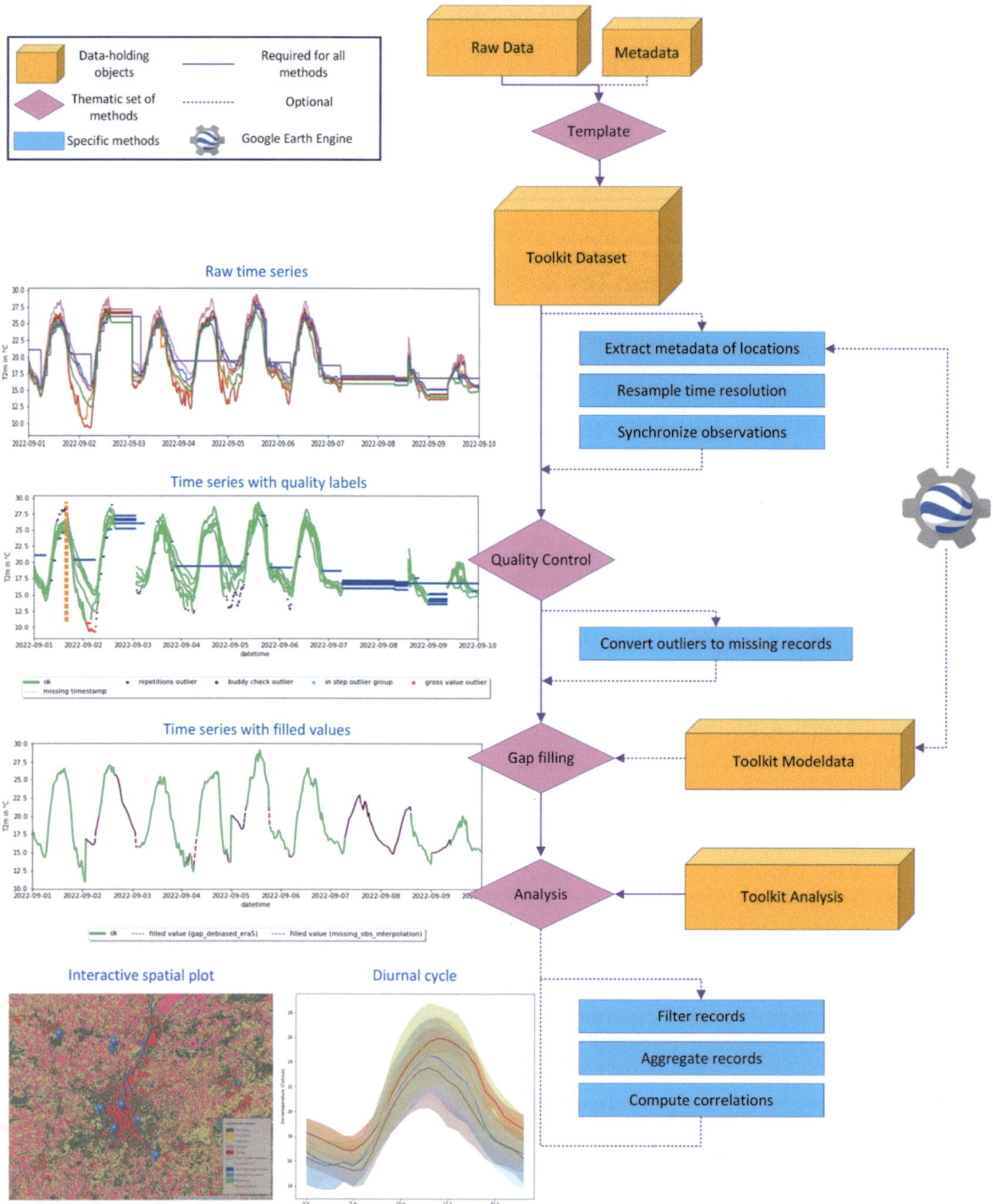

Fig. 4.10 A schematic overview of the main MetObs toolkit functionalities, After Vergauwen et al. (2024), under the Creative Commons Attribution 4.0 International License

- Meteorological services through their growing open data policy (normally quality proofed).
- Increasing number of open-access amateur meteorological stations (no quality proof).
- Commercial (often paid) specialized measurement networks (e.g. in agriculture), (no quality proof).

If such alternative data sources, along with their data quality and representativeness, can

```
[*]: import metobs_toolkit

     #Create a template for the Raw-data
     metobs_toolkit.build_template_prompt()

     This prompt will help to build a template for your data and metadata. Answer the prompt and hit Enter.


      *******      File locations     ***********

      Is there a data file (containing observations)?. (y/n) :  y
     Give the full path to your data file :  /home/thoverga/Documents/dataset_20221004_20221011.csv
      Do you have a file with the metadata?. (y/n) :  n


      *******      Data File     ***********
      ... opening the data file ...
     How is your dataset structured :

        1. Long format (station observations are stacked as rows)
        2. Wide format (columns represent different stations)
        3. Single station format (columns represent observation(s) of one station)
        x. -- not valid --
   (1 - 3) :
```

Fig. 4.11 A snapshot of the built-in dialog assistant for the creation of a template file

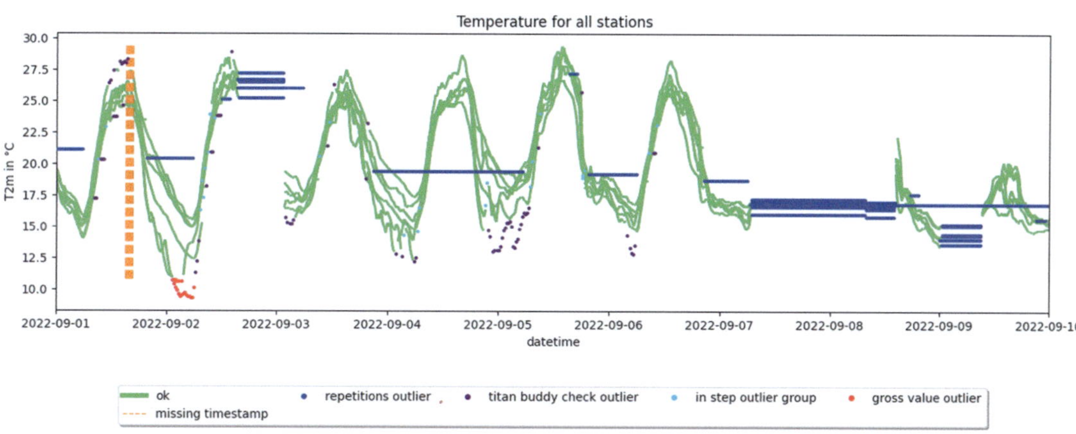

Fig. 4.12 Temperature observations are quality-controlled by a set of specific quality checks. Outliers of these checks are indicated by colour. The green observations are those that passed all checks

be used for the intended application, then the measurement program can be designed by considering already existing and ongoing measurements and available databases to save costs. However, such alternative datasets normally cannot replace accurate micrometeorological measurements, which are required for many applications (such as pest management in agriculture). It must be kept in mind that stations of different networks or at different sites apply various standards, quality control measures, etc., so a careful preliminary check is necessary if these data are in fact useable for the specific purpose.

The Internet provides many of these alternative data sources, such as:

- Databases of international organizations, such as ECMWF, WMO.
- Databases of national meteorological or hydro-meteorological services (e.g. the National Meteorological Service) or organizations, often accessible through a download portal.
- Sites of various agro-meteorological services, where, in addition to irrigation and crop protection information, long station data series are often available.

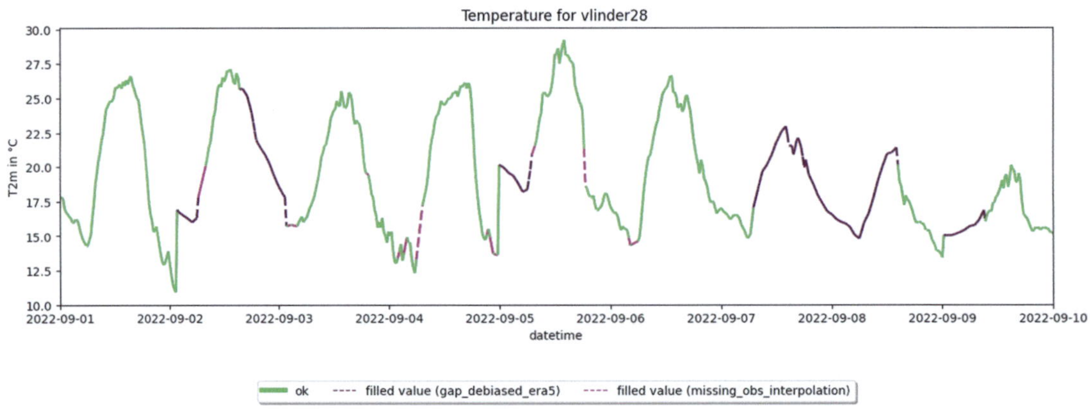

Fig. 4.13 The outliers of the quality control are removed, and the resulting gaps are filled. Short gaps are filled by linear interpolation; larger gaps are filled with debiased ERA5 model data

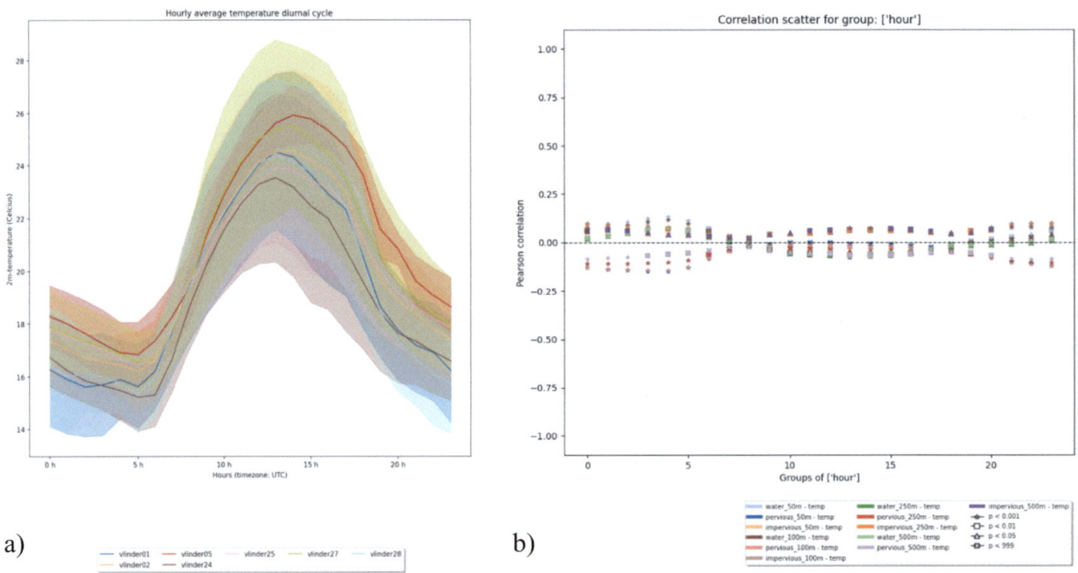

Fig. 4.14 **a** The diurnal cycle of the stations. **b** The diurnal cycle of land cover and temperature (Pearson) correlations. The land cover is quantified by land cover fractions of 100, 250, and 500 m buffers at the stations

- Publicly available urban meteorological networks.
- Micrometeorological measurement networks (e.g. FAIRNESS COST (CA20108) Fair Micromet Portal: ▶ https://fair-micromet.eu/).
- Databases of measurements from private meteorological stations or networks.
- Reanalysis databases. Among these the continuously updated ERA5 (see ▶ Sect. 4.6.2) database with an hourly resolution of approximately 20 km grid resolution is a prominent example (Hersbach et al. 2020). High resolution from 5-km to 1-km grid resolution and from 6 to 1 h time resolution databases have also appeared (Ban et al. 2021; Qin et al. 2022; Thiemig et al. 2022). They mainly contain temperature and precipitation fields.

```
dataset.get_lcz() #get local climate zones for all stations
dataset.get_altitude() #get altitude for all stations

#get landcover fractions for all stations in a buffer of 250m and aggregate
#to impervious/pervious and water classes.
landcover_df = dataset.get_landcover(buffers=[250],
                                     aggregate=True,
                                     gee_map='worldcover')
```

name	lat	lon	geometry	dataset_resolution	lcz	altitude	water_250m	pervious_250m	impervious_250m
vlinder01	50.980438	3.815763	POINT (3.81576 50.98044)	0 days 00:05:00	Low plants (LCZ D)	12	0.000000	0.963635	0.036365
vlinder02	51.022379	3.709695	POINT (3.70969 51.02238)	0 days 00:05:00	Open midrise	7	0.000000	0.535944	0.464056
vlinder05	51.052655	3.675183	POINT (3.67518 51.05266)	0 days 00:05:00	Water (LCZ G)	0	0.242406	0.526977	0.230617
vlinder24	51.167015	3.572062	POINT (3.57206 51.16701)	0 days 00:05:00	Dense Trees (LCZ A)	12	0.000000	0.946138	0.053862
vlinder25	51.154720	3.708611	POINT (3.70861 51.15472)	0 days 00:05:00	Water (LCZ G)	12	0.899936	0.063972	0.036092
vlinder27	51.058099	3.728067	POINT (3.72807 51.05810)	0 days 00:05:00	Compact midrise	12	0.018481	0.084840	0.896679
vlinder28	51.035293	3.769741	POINT (3.76974 51.03529)	0 days 00:05:00	Open lowrise	7	0.000000	0.721950	0.278050

Fig. 4.15 Metadata on the location of the stations is extracted by the use of the Google Earth Engine API

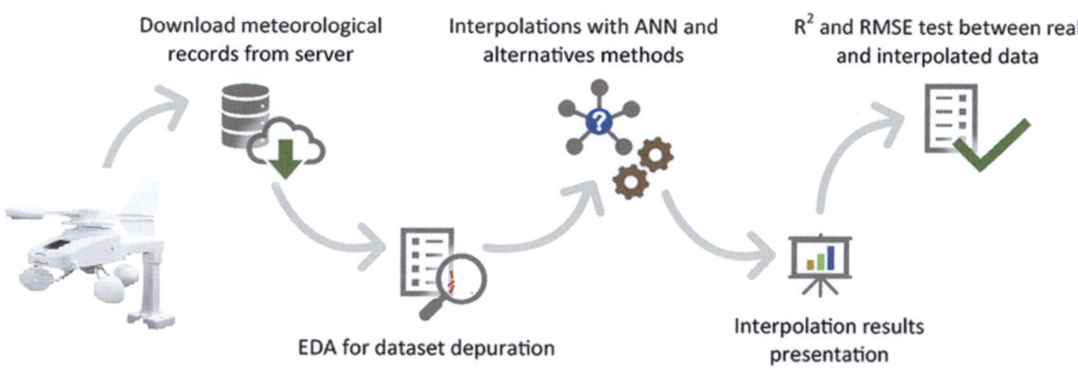

Fig. 4.16 Virtual weather station pseudocode flowchart for obtaining meteorological data, (ANN—Artificial Neural Network, EDA—Exploratory Data Analysis). After Franco et al. (2020), with kind permission of Springer

In that context, a new element of the use of low-cost data is the *virtual weather station/network* (VWS/N) application (Fig. 4.16), which can be installed on a mobile phone, using existing databases (surface and satellite measurements, reanalysis products, models, etc.) that are freely available on the Internet. Here, it is already possible to derive meteorological state variables for a given location using artificial intelligence, deep learning algorithms, and interpolation techniques (Rosillon et al. 2019; Franco et al. 2020).

However, the information obtained from well-estimated in situ measurements exceeds that obtained from interpolated data series at a given location as there can be substantial differences and bias to a specific microclimate. Therefore, these data can provide only a rough estimation on local weather conditions, if no better data sources are available.

They could be used also for a remote check on potential sensor failures at weather stations, e.g. when no rain is recorded due to a collated rain gauge although significant precipitation is shown in the virtual data.

4

4.9 Overview on Existing Meteorological Measurement Standards

The term *standardization* has a broad definition. It ranges from government regulatory activities (legal standards) and public standardization to industry standardization. Public standardization and its products are the results of work performed by standards development organizations (SDOs). Due to their negligible importance to standardization in atmospheric measurements, this section does only address public standards and individual standards of measurement programs but not company, de facto (Jäckel et al. 2021), or application-related standards.

International standards are issued by the International Organization for Standardization (ISO) and European standards by the European Committee for Standardization (CEN). Standards in individual countries are rather rare. In Germany, the Association of German Engineers (VDI 3786, 25 sheets) has standards for measurements of pollutant dispersion and urban climate. Standards in atmospheric measurements usually provide:

- Necessary basics
- Descriptions and/or definitions of the measurement targets
- Specifications for measurements and measurement planning
- Methods of determining equipment characteristics and proper application of measuring equipment including calibration, monitoring, and maintenance
- Requirements for the selection and preparation of a measurement site
- Requirements on data collection, processing, and evaluation.

In the Appendix, ◘ Table 4.6 contains a compilation of the most important international standards that have significance in the context of this book. In addition, two industry standards for temperature sensors are given. Standards for the measurement of various trace gases are not listed.

The World Meteorological Organization (WMO) has been accepted by the ISO as an international standardizing body. The WMO and the ISO also agreed on the development of common ISO-WMO standards (see Appendix, ◘ Table 4.6). Thus, many standardization projects in the interest of WMO are developed in close cooperation, for example, through the WMO's Technical Commission on Instruments and Methods of Observation (CIMO). The CIMO's mission is to give advice on good practices for meteorological measurements and observations and to provide best practices, procedures, and basic capabilities of instruments and systems (Jäckel et al. 2021). The guidelines are continually updated, with Volume 1 relevant to the applications in this book (WMO 2024).

The *WMO has defined so-called Primary and Secondary Standards* for the verification of the measuring instruments. The primary standards are only provided at a few centres in the world. Secondary standards should be available in all countries, and some instrument types today often already comply with these standards. For the applications in this book, working standards are sufficient, which can be used for comparison measurements in the laboratory or directly at the measuring site (◘ Table 4.4).

To monitor the state of and changes in the atmosphere, the WMO has installed a system of measurement programs within the Global Observing System (GOS) that target specific meteorological elements or trace gases (Philipona 2021). These measurement programs have developed their own standards for measurement and quality assurance.

Besides the WMO standards, only standardizations of specific ecological measurement programs are well established and harmonized (Schmid and Rebmann 2021), such as the FLUXNET program of measurement stations of energy and carbon dioxide fluxes (Baldocchi et al. 2001), the special standards of the European Integrated Carbon Observation System (ICOS, Rebmann et al. 2018), and the U.S. National Ecological Observatory Network (NEON, Sturtevant et al. 2022).

Table 4.4 Working standards for different meteorological elements (WMO 2024)

Meteorological element	Working standard
Temperature	Well-calibrated reference thermometer
Humidity	Chilled-mirror dew point hygrometer, Assmann psychrometer, salt solutions
Wind	See ISO 16622, ISO 17713-1, IEC 61400-12
Pressure	Electronic barometers, such as silicon diaphragm barometers with long-term stability
Radiation	See ISO 9060
Precipitation	Portable field calibration system, CEN 16469

Appendix

See Tables 4.5 and 4.6.

Table 4.5 Metadata categories used by the WMO WIGOS system based on WMO (2019) (Tordai 2022)

Category	Description
1. Observed variable	Observed meteorological condition, indicator basic characteristics and description of the resulting data series Data/information on spatial representativeness. Biogeophysical processes described by the observation
2. Purpose of observation	Main applications of monitoring. Observation framework program and the monitoring network, the name of the monitoring network, if any
3. Station/platform	The complex facility where the observations are to be made description of the facility, including fixed and mobile stations and remote sensing equipment
4. Environment	Description of the geographical environment of the observation. Provide geographic coordinates of the stations and heights above sea level
5.Instruments and observation methods	Get the details of observation methods, measuring instruments, and characteristics of measuring instruments. Additionally, calibration characteristics of sensors are crucial
6. Sampling	Sampling process and frequency details of time step and time zone (e.g. UTC) setting. Indicate winter/summer time setting; however, it is recommended to stay always with winter time
7. Data processing and transmission	Description of the path of the raw data stream from the generation of the raw data through the conversion into meteorological variables to the transmission of the data to end users
8. Data quality	The data quality conditions and the description of the traceability of the observation
9.Ownership and data policy	Ownership of observations, responsible persons, participants (persons and institutions), data policy (e.g. open or restricted), details or link to the physical place of the data
10. Contact	The information on the monitoring and the contact details, contact persons

4

Table 4.6 Selected standards in atmospheric measurements

No.: Year	Standard	Purpose
ISO 9060:2018	Solar energy—Specification and classification of instruments for measuring hemispherical solar and direct solar radiation	Sensor
ISO 11276:1995	Soil quality—Determination of pore water pressure—Tensiometer method	Sensor, method
ISO 16622:2002	Meteorology—Sonic anemometer/thermometer—Acceptance test method for mean wind measurements	Sensor, calibration
ISO 16586:2006 Corrigendum 2009	Soil quality—Determination of soil water content as a volume fraction on the basis of known dry bulk density—Gravimetric method	Sensor, method
ISO 17713-1:2007	Meteorology—Wind measurements—Part 1: Wind tunnel test methods for rotating anemometer performance	Sensor, calibration
ISO 17714:2007	Meteorology—Air temperature measurements—Test methods for comparing the performance of thermometer shields/screens and defining important characteristics	Sensor, characteristics
ISO 19289:2015	Air quality—Meteorology—Siting classifications for surface observing stations on land (aligned with WMO 2024, Annex 1D)	Siting
ISO 28902-1:2012	Air quality—Environmental meteorology—Part 1: Ground-based remote sensing of visual range by lidar	Sensor, method
ISO 28902-2:2016	Air quality—Environmental meteorology—Part 2: Ground-based remote sensing of wind by heterodyne pulsed Doppler lidar	Sensor, method
CEN 16469:2013	Hydrometry—Measurement of rainfall intensity (Liquid precipitation): requirements, calibration methods, and field measurements	Calibration
IEC 60751:2008[a] DIN-EN 60751:2009[b]	Industrial platinum resistance thermo0meter, Edition 2.0	Sensor element
IEC 61400-12:2022[a]	Wind energy generation systems—Part 12: Power performance measurements of electricity producing wind turbines—Overview	Sensor
IEC 62460:2008[a] DIN-EN 62460:2009[b]	Temperature—Electromotive force (EMF) tables for pure-element thermocouple combinations	Sensor element

[a] International Electrotechnical Commission
[b] German Industrial Norm—European Norm

References

Aguilar E, Auer I, Brunet M, Peterson TC, Wieringa J (2003) Guidelines on climate metadata and homogenization, WMO/TD-No. 1186. World Meteorological Organization, Geneva

Baldocchi D, Falge E, Gu L, Olson R, Hollinger D, Running S, Anthoni P, Bernhofer C, Davis K, Evans R, Fuentes J, Goldstein A, Katul G, Law B, Lee XH, Malhi Y, Meyers T, Munger W, Oechel W, Paw UKT, Pilegaard K, Schmid HP, Valentini R, Verma S, Vesala T (2001) FLUXNET: A new tool to study the temporal and spatial variability of ecosystem-scale carbon dioxide, water vapor, and energy flux densities. Bull Amer Meteorol Soc 82:2415–2434. ► https://doi.org/10.1175/1520-0477(2001)082%3c2415:FANTTS%3e2.3.CO;2

Ban N, Caillaud C, Coppola E, Pichelli E, Sobolowski S, Adinolfi M, Ahrens B, Alias A, Anders I, Bastin S, Belušić D, Berthou S, Brisson E, Cardoso RM, Chan SC, Christensen OB, Fernández J, Fita L, Frisius T, Gašparac G, Giorgi F, Goergen K, Haugen JE, Hodnebrog Ø, Kartsios S, Katragkou E, Kendon EJ, Keuler K, Lavin-Gullon A, Lenderink G, Leutwyler D, Lorenz T, Maraun D, Mercogliano P, Milovac J, Panitz H-J, Raffa M, Remedio AR,

Schär C, Soares PMM, Srnec L, Steensen BM, Stocchi P, Tölle MH, Truhetz H, Vergara-Temprado J, de Vries H, Warrach-Sagi K, Wulfmeyer V, Zander MJ (2021) The first multi-model ensemble of regional climate simulations at kilometer-scale resolution, part I: evaluation of precipitation. Clim Dyn 57:275–302. ► https://doi.org/10.1007/s00382-021-05708-w

Beckers J-M, Rixen M (2003) EOF calculations and data filling from incomplete oceanographic datasets. J Atmos Oceanic Techn 20:1839–1856. ► https://doi.org/10.1175/1520-0426(2003)020%3c1839:ECADFF%3e2.0.CO;2

Blyth E, Gash J, Lloyd A, Pryor M, Weedon GP, Shuttleworth J (2010) Evaluating the JULES land surface model energy fluxes using FLUXNET data. J Hydrometeorol 11:509–519. ► https://doi.org/10.1175/2009JHM1183.1

Brandsma T, Buishand TA (1998) Simulation of extreme precipitation in the Rhine basin by nearest-neighbour resampling. Hydrol Earth Sys Sci 2:195–209. ► https://doi.org/10.5194/hess-2-195-1998

Brock FV (1986) A nonlinear filter to remove impulse noise from meteorological data. J Atmos Oceanic Techn 3:51–58. ► https://doi.org/10.1175/1520-0426(1986)003%3c0051:ANFTRI%3e2.0.CO;2

Brunet M, Brugnara Y, Noone S, Stephens A, Valente MA, Ventura C, Jones P, Gilabert A, Brönnimann S, Luterbacher J (2020) Best practice guidelines for climate data and metadata formatting, quality control and submission. Quality Control and Submission of the Copernicus Climate Change Service Data Rescue Service. ► https://doi.org/10.24381/kctk-8j22

Cerlini PB, Silvestri L, Saraceni M (2020) Quality control and gap-filling methods applied to hourly temperature observations over central Italy. Meteorol Appl 27:e1913. ► https://doi.org/10.1002/met.1913

Claridge DE, Chen H (2006) Missing data estimation for 1–6 h gaps in energy use and weather data using different statistical methods. Int J Energy Res 30:1075–1091. ► https://doi.org/10.1002/er.1207

Daly C, Gibson WP, Taylor GH, Johnson GL, Pasteris P (2002) A knowledge-based approach to the statistical mapping of climate. Climate Res 22:99–113. ► https://doi.org/10.3354/cr022099

Daly SF, Davis R, Ochs E, Pangburn T (2000) An approach to spatially distributed snow modelling of the Sacramento and San Joaquin basins, California. Hydrol Process 14:3257–3271. ► https://doi.org/10.1002/1099-1085(20001230)14:18%3c3257::AID-HYP199%3e3.0.CO;2-Z

Dodson R, Marks D (1997) Daily air temperature interpolated at high spatial resolution over a large mountainous region. Climate Res 8:1–20. ► https://doi.org/10.3354/cr008001

Domonkos P, Tóth R, Nyitrai L (2023) Climate observations: data quality control and time series homogenization. Elsevier, Amsterdam. ► https://doi.org/10.1016/C2020-0-03312-8

Dumitrescu A, Brabec M, Cheval S (2020) Statistical gap-filling of SEVIRI land surface temperature. Remote Sens 12:1423. ► https://doi.org/10.3390/rs12091423

Dyukarev E (2023) Comparison of artificial neural network and regression models for filling temporal gaps of meteorological variables time series. Appl Sci 13:2646. ► https://doi.org/10.3390/app13042646

El Hachimi C, Belaqziz S, Khabba S, Ousanouan Y, Sebbar B, Kharrou MH, Chehbouni A (2023) ClimateFiller: a python framework for climate time series gap-filling and diagnosis based on artificial intelligence and multi-source reanalysis data. Softw Impacts 18:100575. ► https://doi.org/10.1016/j.simpa.2023.100575

Fagandini C, Todaro V, Tanda MG, Pereira JL, Azevedo L, Zanini A (2024) Missing rainfall daily data: a comparison among gap-filling approaches. Math Geosci 56:191–217. ► https://doi.org/10.1007/s11004-023-10078-6

Finkelstein PL, Sims PF (2001) Sampling error in eddy correlation flux measurements. J Geophys Res D106:3503–3509. ► https://doi.org/10.1029/2000JD900731

Franco BM, Hernández-Callejo L, Navas-Gracia LM (2020) Virtual weather stations for meteorological data estimations. Neural Comput Appl 32:12801–12812. ► https://doi.org/10.1007/s00521-020-04727-8

Garen DC, Johnson GL, Hanson CL (1994) Mean Areal precipitation for daily hydrologic modeling in mountainous regions. JAWRA J Am Water Res Assoc 30:481–491. ► https://doi.org/10.1111/j.1752-1688.1994.tb03307.x

Guijarro JA, López JA, Aguilar E, Domonkos P, Venema VKC, Sigró J, Brunet M (2023) Homogenization of monthly series of temperature and precipitation: benchmarking results of the MULTITEST project. Int J Climatol 43:3994–4012. ► https://doi.org/10.1002/joc.8069

Hartkamp AD, De Beurs K, Stein A, White JW (1999) Interpolation techniques for climate variables. NRG-GIS Series 99-01. CIMMYT, Mexico, D.F.

Henn B, Raleigh MS, Fisher A, Lundquist JD (2013) A comparison of methods for filling gaps in hourly near-surface air temperature data. J Hydrometeorol 14:929–945. ► https://doi.org/10.1175/JHM-D-12-027.1

Hersbach H, Bell B, Berrisford P, Hirahara S, Horányi A, Muñoz-Sabater J, Nicolas J, Peubey C, Radu R, Schepers D, Simmons A, Soci C, Abdalla S, Abellan X, Balsamo G, Bechtold P, Biavati G, Bidlot J, Bonavita M, De Chiara G, Dahlgren P, Dee D, Diamantakis M, Dragani R, Flemming J, Forbes R, Fuentes M, Geer A, Haimberger L, Healy S, Hogan RJ, Hólm E, Janisková M, Keeley S, Laloyaux P, Lopez P, Lupu C, Radnoti G, De Rosnay P, Rozum I, Vamborg F, Villaume S, Thépaut J-N (2020) The ERA5 global reanalysis. Quart J Roy Meteorol Soc 146:1999–2049. ► https://doi.org/10.1002/qj.3803. doi:10.1002/qj.3803

Jacobs A, Top S, Vergauwen T, Suomi J, Käyhkö J, Caluwaerts S (2024) Filling gaps in urban temperature observations by debiasing ERA5 reanalysis data.

Urban Clim 58:102226. ► https://doi.org/10.1016/j.uclim.2024.102226

Jäckel S, Borowiak A, Stacey B (2021) Standardization in atmospheric measurements. In: Foken T (Ed) Springer handbook of atmospheric measurements. Springer International Publishing, Cham, pp 93–106. ► https://doi.org/10.1007/978-3-030-52171-4_4

Körner P, Kronenberg R, Genzel S, Bernhofer C (2018) Introducing gradient boosting as a universal gap filling tool for meteorological time series. Meteorol Z 27:369–376. ► https://doi.org/10.1127/metz/2018/0908

Liston GE, Elder K (2006) A meteorological distribution system for high-resolution terrestrial modeling (MicroMet). J Hydrometeorol 7:217–234. ► https://doi.org/10.1175/JHM486.1

Lompar M, Lalić B, Dekić L, Petrić M (2019) Filling gaps in hourly air temperature data using debiased ERA5 data. Atmosphere 10:13. ► https://doi.org/10.3390/atmos10010013

Metzger S, Junkermann W, Mauder M, Beyrich F, Butterbach-Bahl K, Schmid HP, Foken T (2012) Eddy-covariance flux measurements with a weight-shift microlight aircraft. Atmos Meas Techn 5:1699–1717. ► https://doi.org/10.5194/amt-5-1699-2012

Minder JR, Mote PW, Lundquist JD (2010) Surface temperature lapse rates over complex terrain: lessons from the Cascade mountains. J Geophys Res: Atmos 115(D14). ► https://doi.org/10.1029/2009JD013493

Papale D (2012) Data gap filling. In: Aubinet M, Vesala T, Papale D (Eds) Eddy covariance: a practical guide to measurement and data analysis. Springer, Dordrecht, Heidelberg, London, New York, pp 159–172. ► https://doi.org/10.1007/978-94-007-2351-1_6

Pape R, Wundram D, Löffler J (2009) Modelling near-surface temperature conditions in high mountain environments: an appraisal. Climate Res 39:99–109. ► https://doi.org/10.3354/cr00795

Pastorello G, Agarwal D, Papale D, Samak T, Trotta C, Ribeca A, Poindexter C, Faybishenko B, Gunter D, Hollowgrass R (2014) Observational data patterns for time series data quality assessment. In: 2014 IEEE 10th international conference on e-science. IEEE, pp 271–278. ► https://doi.org/10.1109/eScience.2014.45

Petrić M, Marsboom C, Vandendriessche J (2020) Wireless sensor networks in IPM. In: Social, legal, and ethical implications of IoT, cloud, and edge computing technologies. IGI Global, pp 28–52. ► https://doi.org/10.4018/978-1-7998-3817-3.ch00

Philipona R (2021) Networks of atmospheric measuring techniques. In: Foken T (ed) Springer handbook of atmospheric measurements. Springer, Cham, pp 1687–1712. ► https://doi.org/10.1007/978-3-030-52171-4_63

Qin J, He M, Jiang H, Lu N (2022) Reconstruction of 60-year (1961–2020) surface air temperature on the Tibetan Plateau by fusing MODIS and ERA5 temperatures. Sci Total Environ 853:158406. ► https://doi.org/10.1016/j.scitotenv.2022.158406

R Core Team (2024) R: a language and environment for statistical computing. R Foundation for Statistical Computing, Vienna, Austria. ► https://www.R-project.org/

Rebmann C, Aubinet M, Schmid HP, Arriga N, Aurela M, Burba G, Clement R, De Ligne A, Fratini G, Gielen B, Grace J, Graf A, Gross P, Haapanala S, Herbst M, Hörtnagl L, Ibrom A, Joly L, Kljun N, Kolle O, Kowalski A, Lindroth A, Loustau D, Mammarella I, Mauder M, Merbold L, Metzger S, Mölder M, Montagnani L, Papale D, Pavelka M, Peichl M, Roland M, Serrano-Ortiz P, Siebicke L, Steinbrecher R, Tuovinen J-P, Vesala T, Wohlfahrt G, Franz D (2018) ICOS eddy covariance flux-station site setup: a review. Int Agrophys 32:471–494. ► https://doi.org/10.1515/intag-2017-0044

Rolland C (2003) Spatial and seasonal variations of air temperature lapse rates in Alpine regions. J Climate 16:1032–1046. ► https://doi.org/10.1175/1520-0442(2003)016<1032:SASVOA>2.0.CO;2

Rosillon D, Huart JP, Goossens T, Journée M, Planchon V (2019) The agromet project: a virtual weather station network for agricultural decision support systems in Wallonia, South of Belgium. In: Palattella M, Scanzio S, Coleri Ergen S (eds) Ad-hoc, mobile, and wireless networks. ADHOC-NOW 2019. Lecture notes in computer science 11803. Springer, Cham. ► https://doi.org/10.1007/978-3-030-31831-4_39

Schmid HP, Rebmann C (2021) Integration of meteorological and ecological measurements. In: Foken T (Ed) Springer handbook of atmospheric measurements. Springer International Publishing, Cham, pp 1713–1725. ► https://doi.org/10.1007/978-3-030-52171-4_64

Sibson R (1981) A brief description of natural neighbour interpolation. In: Barnett V (ed) Interpreting multivariate data. Wiley, New York, pp 21–36

Stahl K, Moore RD, Floyer JA, Asplin MG, McKendry IG (2006) Comparison of approaches for spatial interpolation of daily air temperature in a large region with complex topography and highly variable station density. Agric Forest Meteorol 139:224–236. ► https://doi.org/10.1016/j.agrformet.2006.07.004

Starkenburg D, Metzger S, Fochesatto GJ, Alfieri JG, Gens R, Prakash A, Cristóbal J (2016) Assessment of despiking methods for turbulence data in micrometeorology. J Atm Oceanic Techn 33:2001–2013. ► https://doi.org/10.1175/JTECH-D-15-0154.1

Stöckli R, Lawrence DM, Niu G-Y, Oleson KW, Thornton PE, Yang Z-L, Bonan GB, Denning AS, Running SW (2008) Use of FLUXNET in the community land model development. J Geophys Res: Biogeosci 113(G1). ► https://doi.org/10.1029/2007JG000562

Sturtevant C, Metzger S, Nehr S, Foken T (2021) Quality assurance and control. In: Foken T (ed) Springer handbook of atmospheric measurements. Springer Nature, Cham, pp 49–92. ► https://doi.org/10.1007/978-3-030-52171-4_3

Sturtevant C, DeRego E, Metzger S, Ayres E, Allen D, Burlingame T, Catolico N, Cawley K, Csavina J, Durden D, Florian C, Frost S, Gaddie R, Knapp E, Laney C, Lee R, Lenz D, Litt G, Luo H, Roberti J, Slemmons C, Styers K, Tran C, Vance T, San-Clements M (2022) A process approach to quality management doubles NEON sensor data quality. Methods Ecol Evol 13:1849–1865. ► https://doi.org/10.1111/2041-210X.13943

Taylor JR, Loescher HL (2013) Automated quality control methods for sensor data: a novel observatory approach. Biogeoscience 10:4957–4971. ► https://doi.org/10.5194/bg-10-4957-2013

Thiemig V, Gomes GN, Skøien JO, Ziese M, Rauthe-Schöch A, Rustemeier E, Rehfeldt K, Walawender JP, Kolbe C, Pichon D, Schweim C (2022) Salamon P (2021) EMO-5: a high-resolution multi-variable gridded meteorological dataset for Europe. Earth Syst Sci Data 14:3249–3272. ► https://doi.org/10.5194/essd-14-3249-2022

Thiessen AH (1911) Precipitation averages for large areas. Mon Weather Rev 39:1082–1089. ► https://doi.org/10.1175/1520-0493(1911)39%3c1082b:PAFLA%3e2.0.CO;2

Tobin C, Nicotina L, Parlange MB, Berne A, Rinaldo A (2011) Improved interpolation of meteorological forcings for hydrologic applications in a Swiss Alpine region. J Hydrol 401:77–89. ► https://doi.org/10.1016/j.jhydrol.2011.02.010

Tordai AV (2022) Micrometeorological measurement programs (2015–2021) building a unified database. MSc Thesis in Department of Meteorology Eötvös Loránd University, 92 p. ► https://nimbus.elte.hu/tanszek/docs/MSc/2022_2/Tordai_Agoston_Vilmos_2022.pdf (In Hung.)

Vergauwen T, Vieijra M, Covaci A, Jacobs A, Top S, Dewettinck W, Vandelanotte K, Hellebosch I, Caluwaerts S (2024) MetObs—A Python toolkit for using non-traditional meteorological observations. JOSS 9:5916. ► https://doi.org/10.21105/joss.05916

Venema V, Trewin B, Wang XL, Szentimrey T, Lakatos M, Aguilar E, Auer I, Guijarro J, Menne M, Oria C, Louamba WSRL, Rasul G (2020) Guidelines on homogenization. World Meteorological Organization, WMO-No. 1245, Geneva, Switzerland

Vickers D, Mahrt L (1997) Quality control and flux sampling problems for tower and aircraft data. J Atmos Oceanic Technol 14:512–526. ► https://doi.org/10.1175/1520-0426(1997)014%3c0512:-QCAFSP%3e2.0.CO;2

Von Storch H, Zwiers FW (1999) Statistical aanalysis in climate research. Cambridge University Press. ► https://doi.org/10.1017/CBO9780511612336

Vuichard N, Papale D (2015) Filling the gaps in meteorological continuous data measured at FLUXNET sites with ERA-Interim reanalysis. Earth Syst Sci Data 7:157–171. ► https://doi.org/10.5194/essd-7-157-2015

WMO (2019) WIGOS Metadata Standard (WMO-No. 1192). World Meteorological Organisation, Geneva

WMO (2024) Guide to instruments and methods of observation, WMO-No. 8, Volume I—Measurement of meteorological variables. World Meteorological Organization, Geneva

WMO (2023a) Guide to the WMO integrated global observing system, WMO-No. 1165. World Meteorological Organisation, Geneva

WMO (2023b) Manual on the WMO integrated global observing system, WMO-No. 1160. World Meteorological Organisation, Geneva

Zeng ML, Qin J (2020) Metadata, 3rd edn. Facet Publishing, London, p 592

Supplementary Information

B. Lalic et al. (eds.), *Micrometeorological Measurements*, Springer Textbooks in Earth Sciences, Geography and Environment, https://doi.org/10.1007/978-3-032-03884-5

Index

I

L

M

N

O

P

Q

R

MIX
Papier aus verantwortungsvollen Quellen
Paper from responsible sources
FSC® C105338

If you have any concerns about our products,
you can contact us on
ProductSafety@springernature.com

In case Publisher is established outside the EU,
the EU authorized representative is:
Springer Nature Customer Service Center GmbH
Europaplatz 3, 69115 Heidelberg, Germany

Printed by Libri Plureos GmbH
in Hamburg, Germany